Mechanics of Materials

Parviz Ghavami

Mechanics of Materials

An Introduction to Engineering Technology

 Springer

Parviz Ghavami
Harlingen, TX, USA

ISBN 978-3-319-36044-7   ISBN 978-3-319-07572-3 (eBook)
DOI 10.1007/978-3-319-07572-3
Springer Cham Heidelberg New York Dordrecht London

Printed on acid-free paper

Springer is part of Springer Science+Business Media (www.springer.com)

Preface

There has been a need for a textbook in mechanics of materials for students in 2- or 4-year technology programs in engineering, architecture, or building construction. In addition, students in vocational schools and technical institutes will find in this book the fundamentals of statics and strength of materials that will be vital in school and in their professional lives.

This text provides the necessary information required for students of the aforementioned programs to successfully grasp important physical concepts. The material has been written in a simple way, with only some knowledge of college algebra and trigonometry required; no special engineering background or knowledge of calculus is necessary for understanding this text.

I have taught statics and strength of materials in the engineering program at Texas State Technical College-Harlingen for the past 15 years, and I have presented the material here in the same way that I would present it in the classroom. Each topic is followed by examples so the student can learn the problem solving methods and apply them in real-world problems. The student will also see a set of practice problems at the end of each topic. At the end of each chapter, there is a summary with a set of review questions and problems.

Doing the homework will give the student a much deeper understanding of the variety of concepts and encourage him/her to continue studying this fascinating branch of engineering. It goes without saying that the material in this text could also be a valuable reference for those individuals seeking state licensing in professional engineering. The students taking this class must not just read the book; they should take it as a serious and important text, concentrating on each chapter, working on the problems carefully, analyzing each problem, and trying to relate them to real-world situations or problems they are currently facing on the job site. This is how to learn engineering science. They always say that mastery of technical ideas often means hard work and concentration. Those who are not afraid of a challenge can excel!

I have organized the chapters in a simple way so the student can easily read and understand the material. I have avoided using difficult language. In fact, the book is based on the simplest educational process; I believe that in writing such a technical book this method must be followed. In Chap. 1, I introduce the basic fundamentals, definitions in mechanics of materials, and the metric and English systems of units.

I have shown examples and also included some questions and problems for the student to work on. Working on engineering problems requires a firm understanding of how to do unit conversions, because the unit conversion is the building block of engineering science. In Chap. 2, I cover force systems on structures, force components, equilibrium of forces, and the free body diagram.

Chapter 3 discusses force moments, principle of moments, moment equations for equilibrium, and application of force moments in engineering. Numerous example problems are introduced and solved throughout this chapter. Chapter 4 covers the centroid of an area, and Chap. 5 discusses the moment of inertia of an area, and many example problems are solved throughout this chapter to clarify this important concept.

Stress and strain are introduced in Chap. 6, and numerous real-world problems are solved in all the sections of the chapter. Chapter 7 covers torsion in circular sections, and discusses how to calculate the transmission of power through a rotating shaft. In Chap. 8, I explain the shear and bending moment in a beam and show many example problems for which the student might find applications in structural engineering and building construction.

Chapter 9 discusses bending stresses in beams and covers the resisting moment and flexure formulas for beams. In Chap. 10, I explain the columns and slenderness ratio for compression members. Steel and timber columns are also discussed in this chapter with example problems.

There are two appendices in this book, A and B. Appendix A shows the beam diagrams and formulas for helping the students to solve the homework problems. Appendix B provides information about the centroids and properties of areas. The students are encouraged to refer to any updated strength of materials or engineering texts to extract more information about the standard steel or timber, if necessary. Author decided not to include the tables of properties from the other current sources, because they may lack data or may be outdated.

In closing, my academic experience teaching mathematics, physics, and engineering courses for 27 years at colleges and universities in Texas, and also my practice as a professional engineer were the primary impetus for writing this book. This book will hopefully show how the fundamental concepts of mechanics of materials can be applied to real-world problems.

I wish to extend my thanks to the staff and faculty at Texas State Technical College-Harlingen, who provided me guidance, encouragement, and support. Finally, I am grateful to my loving son Reza, for his help and encouragement in making this book possible.

I have tried to produce an error-free book, but no doubt some errors still remain. Please let me know of any that you find. Comments, suggestions, and criticisms are always welcome from readers.

Harlingen, TX, USA   Parviz Ghavami

Contents

Author Bios

Parviz Ghavami was born on January 10, 1943 in Iran. He obtained his early and secondary education in Esphahan. He continued his education towards a university degree and received a Master of Science degree in mechanical engineering. He stayed in the engineering field from 1965 to 1978, and worked as a design engineer and senior project engineer for overseas industries both in Iran and abroad.

Leaving Iran in 1978 to pursue further graduate studies in The United States, Dr. Ghavami obtained a Master of Science degree in mechanical engineering at the University of Portland, OR in August 1979. From 1979 until 1983, he worked as a project design engineer for consulting firms in Fort Collins, CO, and Norman, OK.

In August 1984 he joined the faculty of the Mathematics and Science Department at Texas State Technical College in Harlingen, where he taught mathematics, physics, and engineering courses. Meanwhile, in 1989, he accepted an adjunct faculty position with the University of Texas at Brownsville, TX where he taught mathematics, physics, and engineering courses.

In May 1997, he received his Master of Science degree in mathematics at the University of Texas-Pan American. In May 2000, Dr. Ghavami was licensed by the Texas Board of Professional Engineers with a specialization in mechanical and civil engineering.

In May 2003, Dr. Ghavami received a Doctor of Education degree in Administration and Supervision at the University of Houston. Finally, he received his Master of Science degree in civil engineering at Texas A&M University at Kingsville in May 2011.

In the last 15 years, Dr. Ghavami, as president of Ghavami Consulting, has done a great number of projects for the construction industry in residential and commercial structural and mechanical engineering design and construction.

Dr. Ghavami enjoys traveling around the world. His hobbies are reading, gardening, listening to music, cooking, and translating science/science fiction books. In addition to his native language, Farsi, Dr. Ghavami also reads, writes, and speaks German and Russian.

Introduction 1

Overview

The importance of a thorough knowledge of fundamentals in any field cannot be overemphasized. Fundamentals have always been stressed in the learning of new skills, whether it be football or physics. Similarly, the science of mechanics is founded on basic concepts and forms the groundwork for further study in the design and analysis of machines and structures.

Learning Objectives

Upon completion of this chapter, you will be able to define the fundamental terms used in mechanics of materials, and the English or metric systems of units for different problems. You will also be able to differentiate vector and scalar quantities and identify the significance rule of these quantities in the field of mechanics of materials. Your knowledge, application, and problem solving skills will be determined by your performance on the chapter test.

Upon completion of this chapter, you will be able to:

- *Define mechanics of materials*
- *Define the fundamental terms used in mechanics*
- *Identify the main differences in the metric and English systems of units*
- *Define vector and scalar quantities with some examples*

1.1 Mechanics of Materials

Mechanics is defined as the study of the effects of forces on bodies. Statics is the study of bodies that are at rest or moving with constant velocity while subjected to force systems. When the changes of shape of the body and the internal state of the body due to the effects of external force systems become important, the study is

© Springer International Publishing Switzerland 2015
P. Ghavami, *Mechanics of Materials*, DOI 10.1007/978-3-319-07572-3_1

then known as mechanics of materials or strength of materials. It is essential that the following basic terms be understood, since they continually recur in all phases of this technical study.

1.2 Trigonometry

To analyze the forces and work on problems in mechanics of materials or any type of engineering problems, the student must have a thorough understanding of algebraic and trigonometric functions and formulas. Solution of mechanics of materials problems requires such mathematical principles.

1.3 Metric and English Systems of Units

Units are used to define the size of physical quantities. Meter, kilogram, second, newton, and Kelvin are, respectively, units of length, mass, time, force (weight), and temperature in the metric system (SI). Foot, slug, second, pound, and degrees Rankin are, respectively, units of length, mass, time, force (weight), and temperature in English system.

The metric system (SI) offers major advantages relative to the English system. For example, the metric system uses only one basic unit for length, the meter, whereas, the English system uses many basic units for length such as inch, foot, yard, mile, etc. Also, because the metric system is based on multiples of 10, it is easier to use and learn.

The metric system of units, today, has been adopted all over the world. However, the United States is making progress toward the adoption of SI units in order to sell American products more easily on the world market. Therefore, information about the conversion factors is provided between the SI and the English system of units.

1.4 Fundamental Terms

Mass
Mass is a measure of the quantity of matter. It is related to the inertia of the body and is usually considered a constant. The unit of mass in the metric system is the kilogram.

Force
This term is applied to any action on a body which tends to make it move, change its motion, or change its size and shape. A force usually produces acceleration. The unit of force in the metric system is the newton (N) and in the English system, the pound (lb).

Example 1.1 Convert 75 lb to newtons.
(75 lb) (4.45 N/1 lb) = 334 N.

Example 1.2 Convert 150,000 newtons to pounds (lb)
(150,000 N) (1 lb/4.45 N) = 3.37 × 10⁴ lb

Pressure

Pressure is the external force per unit area. It is calculated by dividing the total external force acting on a cross-sectional area of a body or substance. The unit of pressure in the metric system is N/m^2 (Pa), and in the English system, $lb/in.^2$ (psi).

Example 1.3 Find the pressure, in metric units, that a hollow cast-iron column exerts on its foundation shown in Fig. 1.1.

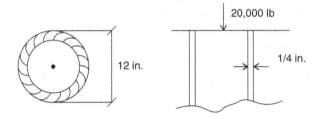

Fig. 1.1

Solution Outside diameter (OD) = 12 in.
Inside diameter (ID) = 12 − 2(0.25) = 11.5 in.
Area of the column $(A) = \pi/4(12^2 - 11.5^2) = 9.23$ in.^{2}
Pressure $(P) =$ Force/Area $= 20,000$ lb/9.23 in.$^2 = 2,167$ lb/in.^{2} (psi)
Using conversion factors: 1 kPa = 1,000 $N/m^2 = 0.145$ lb/in.^{2}
(2,167 psi) (1 kPa/0.145 psi) = 14,944.83 kPa

Density

This term may refer either to weight or mass density. Weight density is the weight per unit volume of a body or substance. The unit of weight density in the metric system is N/m^3 and in the English system, lb/ft^3. Mass density is the mass per unit volume of a body or substance. The unit of mass density in the metric system is kg/m^3 and in the English system, slug/ft^3.

Weight

Weight is the force with which a body is attracted toward the center of the earth by gravitational pull. The unit of weight in the metric system is the newton (N), and in the English system, the pound (lb).

Example 1.4 Find the weight of a wooden block 20 cm × 15 cm × 10 cm. Assume the weight density of wood to be
$D = 227 \text{ N/m}^3$.
Find the volume of the block:
$V = \text{width} \times \text{depth} \times \text{length} = 10 \text{ cm} \times 15 \text{ cm} \times 20 \text{ cm} = 3,000 \text{ cm}^3$
Using conversion factors: $1 \text{ cm}^3 = 10^{-6} \text{ m}^3$
$(3,000 \text{ cm}^3) (10^{-6} \text{ m}^3/1 \text{ cm}^3) = 0.003 \text{ m}^3$
Find the weight of the block:
$W = \text{VD} = 0.003 \text{ m}^3 \times 227 \text{ N/m}^3 = 0.681 \text{ N}.$

Load
This term is used to indicate that a body of some weight is applying a force against some supporting structure or part of a structure. For example, a load weighing 100 lb is applied on a beam supported at two ends. Or, a beam itself can be considered a certain load on part of a structure.

Example 1.5 A brick wall 6 in. thick and 8 ft high supports a roof load equal to 1,500 lb/ft of wall. If the reinforced concrete footing of the wall is 1.7 ft deep and 2.5 ft wide, find the pressure between the footing and the soil (Fig. 1.2) (consider 1 ft of the wall).

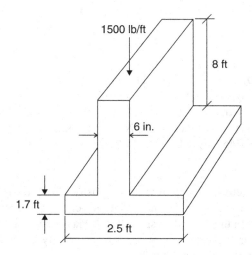

Fig. 1.2

Solution Weight per unit volume (from the table of structural materials) for brick and concrete are:

120 lb/ft^3 (brick)
150 lb/ft^3 (concrete)

Load on one linear foot of wall:

> 1. Roofload $= 1,500$ lb
> 2. Brickwall $= (6/12 \text{ ft})(1 \text{ ft})(8 \text{ ft})(120 \text{ lb/ft}^3)$ $= 480$ lb
> 3. Footing $= (1.7 \text{ ft})(2.5 \text{ ft})(1 \text{ ft})(150 \text{ lb/ft}^3)$ $= 638$ lb
> Total load $\overline{2,618 \text{ lb}}$

$P = \text{force/area} = 2{,}618 \text{ lb}/2.5 \text{ ft} \times 1 \text{ ft} = 1{,}047 \text{ lb/ft}^2$.

Moment

The tendency of a force to cause rotation about an axis through some point is known as *moment*. Moment (M) of a force (F) about a given point (O), is the product of the force and its perpendicular distance r from the line of action between the force and point O.

The point or axis of rotation is called the *center of moments*. The perpendicular distance between the line of action and the center of rotation is called the *moment arm*.

This can be formulated as:

$$M = F \times r$$

Moment of Force $=$ Magnitude of Force $\times$ Moment Arm

The unit of moment in the metric system is N-m, and in the English system, inch-pounds (in.-lb) or foot-pounds (ft-lb).

Example 1.6 A 10 ft beam has a load of 600 lb at a distance of 2 ft from the left end of the beam. Calculate the moment of load about each end point (Fig. 1.3).

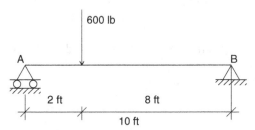

Fig. 1.3

Solution Moment of force $= 600 \text{ lb} \times 2 \text{ ft} = + 1{,}200$ ft-lb
Moment of force $= 600 \text{ lb} \times 8 \text{ ft} = -4{,}800$ ft-lb
Notice the sign of moment clockwise (+) and counterclockwise (−).

Example 1.7 Convert 10.94 in.-lb to newton-meters.
Using conversion factors: 1 lb $= 4.45$ N, and 1 in. $= 0.0254$ m
1 in.-lb $= 0.0254$ m $\times 4.45$ N $= 0.1130$ N-m
Therefore, $(10.94 \text{ in.-lb}) (0.1130 \text{ N-m/in.-lb}) = 1.24$ N-m

Couple

Couple is a pair of parallel forces equal in magnitude and opposite in direction. Their only effect is to produce a moment. The only motion a couple can cause is rotation (Fig. 1.4). Note that the moment of a couple is equal to the product of one of its forces F and the perpendicular distance d between the forces

$$M = F \times d$$

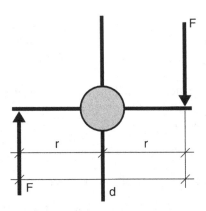

Fig. 1.4

Vector

In general, any quantity that has direction and magnitude is a vector quantity. Examples include force, weight, displacement, velocity, and acceleration, etc.

Scalar

Scalar quantities are quantities that have magnitude only. They are complete without a direction. Examples include mass, density, area, volume, distance, speed, time, temperature, work, power, etc.

1.5   Vector Operations

1.5.1   Multiplication and Division of Vectors by a Scalar

When a vector is multiplied by a scalar quantity, its magnitude will be changed. Depending on the positive or negative values of the scalar, the magnitude of the vector will be increased or decreased. In the same manner, we use this operation if we divide a vector by any positive or negative scalar quantity.

1.5.2   Addition of Vectors

There are two common graphical methods for finding the geometric sum of vectors.

Polygon method
Parallelogram method

Polygon Method

This method is mostly used in applications dealing with the addition of more than two vectors. Use a ruler and protractor to measure the size (magnitude) and direction of the vector. Measurements must be done to proper scale. Continue this process for each vector until you find the magnitude and direction of the resultant vector (Fig. 1.5).

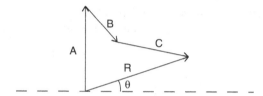

Fig. 1.5

Note that the resultant vector will be drawn with its tail at the origin (starting point) and its tip joined to the tip of the last vector. Also, the order in which the vectors are added together has no effect in obtaining the resultant of the vector.

Parallelogram Method

In the parallelogram method, the resultant of only two forces will be obtained, and the lines of actions of these two forces pass through a common origin. The two forces form the sides of a parallelogram whose diagonal will be represented as the resultant of the two forces (Fig. 1.6). In the parallelogram method, vectors **A** and **B** do not depend upon the order in which they are added. The addition of two vectors is *commutative*, and we write

$$\mathbf{A} + \mathbf{B} = \mathbf{B} + \mathbf{A}$$

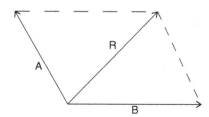

Fig. 1.6

1.5.3 Subtraction of Vectors

Subtraction of vectors **A** and **B** is obtained by adding to **A** the negative vector **B** (−**B**). We write

$$A - B = A + (-B)$$

In graphical representation, **A** − **B** is constructed by connecting the tail of (**A**) to the head of (−**B**).

Subtraction is a special case of addition; therefore, vector addition rules can be applied to vector subtraction.

Example 1.8 A car is pulled out by two cables as shown (Fig. 1.7). If cable A is exerting 1,200 lb, find the force exerted by cable B needed to move the car straight out. Use trigonometric laws to work on the problem.

(Assume that the resultant of forces is directed along the axis of the car.)

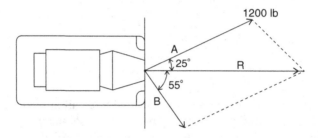

Fig. 1.7

Solution Use a triangle rule. Note that the triangle rule, in fact, is half of the parallelogram rule mentioned earlier. Notice that the resultant of the forces must be perpendicular to the front of the car to keep the car straight.

Using the triangle rule, the force on cable B can be calculated using the law of sines. We write

$$1,200\,\text{lb}/\text{sine}\,55° = B/\sin 25°$$
$$B = 619\ \text{lb}$$

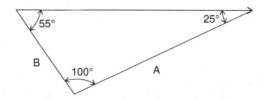

Chapter Summary

1. The subject of mechanics of materials is concerned with the behavior of deformable bodies under the influence of external loads.
2. Fundamental terms such as mass, force, weight, load, density, pressure, moment, and couple are used in our subject of interest.
3. We presented the metric (SI) and English systems of units both used in engineering problems.

   You have learned the major distinction between metric and English systems of units and the advantages of the metric over the English system in our technological world.
4. The concepts of vector and scalar quantities, and how they are applied in mechanics of materials were covered. Vectors have magnitude and direction, whereas scalars are only identified by magnitude.
5. Since the vectors have magnitude and direction, they may not be added in the usual manner. For this purpose, both the polygon and parallelogram methods were presented.

Review Questions

1. What is *mechanics of materials*?
2. What is a *force*?
3. What is a *load*?
4. What is the *moment of a force*?
5. What are the major advantages of the metric system?
6. What is a major distinction between *scalar* and *vector* quantities?
7. Name some examples of scalar and vector quantities in mechanics of materials.
8. Can vectors be added or subtracted the same way as scalars? Why?
9. What method can be applied to add a number of vectors?
10. What method can only be used for the addition of two vectors?

Problems

1. Convert 12.3 ft to SI units.
2. Convert 25.6 lb to kilograms.
3. Convert 875 in.-lb to newton-meters.
4. Convert 6,000 psi to newtons per square meter.
5. Convert 3.45 lb/ft^2 to Pascals (N/m^2).
6. Convert 950 N.m to foot-pounds.
7. Convert 3.5 $in.^3$ to mm^3.
8. Convert 560 ft^2 to m^2.
9. Convert lb/ft to N/m.
10. Convert 1.6×10^9 kPa to MPa.

11. Find the moment of the forces in Fig. 1.8 about the given points:
    (a) 50,000-N force about point B,
    (b) 8,000-N force about point A, and
    (c) 10,000-N force about point A.

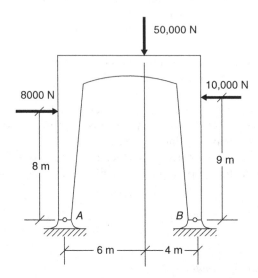

Fig. 1.8

12. State which of the following are scalars and which are vectors.
    (a) Weight
    (b) Energy
    (c) Volume
    (d) Speed
    (e) Momentum
    (f) Distance
13. Represent graphically (a) a force of 10 lb in a direction 30° north of east, (b) a force of 50 N in the direction 60° east of north.
14. Find the magnitude and direction of the resultant of the vectors **A** and **B** which are at right angles. (hint: use the Pythagorean theorem.)
    (a) $A = 15$ N, $B = 20$ N
    (b) $A = 150$ lb, $B = 250$ lb
    (c) $A = 1,500$ N, $B = 3,500$ N
15. Add the given vectors in Fig. 1.9 by drawing the appropriate resultant. Use the parallelogram in (c) and (d).

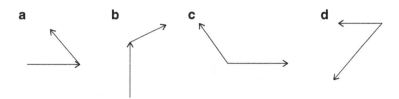

Fig. 1.9

16. Using the polygon method, add the vectors in Fig. 1.10 (choose appropriate scale).

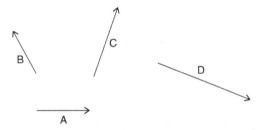

Fig. 1.10

17. Using the parallelogram law and also trigonometric rules, find the resultant of the forces on the structure shown (Fig. 1.11).

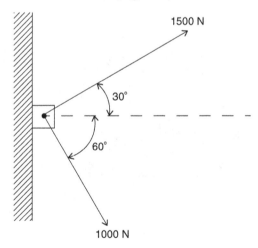

Fig. 1.11

18. Find the resultant of forces on the hook shown (Fig. 1.12) (hint: use the law of cosines.)

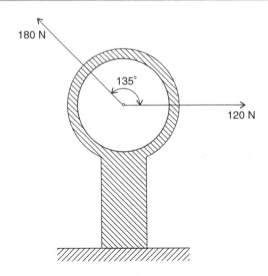

Fig. 1.12

19. Using the component method, find the resultant of the concurrent forces shown
    in Fig. 1.13 on the structural members A, B, and C.

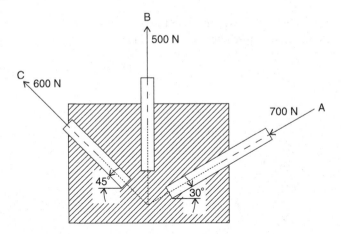

Fig. 1.13

20. If an 800-N force is required to remove a nail from piece of wood, find the horizontal and vertical components of the force (Fig. 1.14).

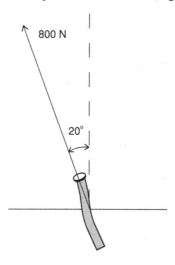

Fig. 1.14

SI prefixes

Multiplication factor	Prefix[a]	Symbol
$1\ 000\ 000\ 000\ 000 = 10^{12}$	Tera	T
$1\ 000\ 000\ 000 = 10^{9}$	Giga	G
$1\ 000\ 000 = 10^{6}$	Mega	M
$1\ 000 = 10^{3}$	Kilo	k
$100 = 10^{2}$	Hecto[b]	h
$10 = 10^{1}$	Deka[b]	da
$0.1 = 10^{-1}$	Deci[b]	d
$0.01 = 10^{-2}$	Centi[b]	c
$0.001 = 10^{-3}$	Milli	m
$0.000\ 001 = 10^{-6}$	Micro	μ
$0.000\ 000\ 001 = 10^{-9}$	Nano	n
$0.000\ 000\ 000\ 001 = 10^{-12}$	Pico	p
$0.000\ 000\ 000\ 000\ 001 = 10^{-15}$	Femto	f
$0.000\ 000\ 000\ 000\ 000\ 001 = 10^{-18}$	Atto	a

[a]The first syllable of every prefix is accented so that the prefix will retain its identity. Thus, the preferred pronunciation of kilometer places the accent on the first syllable, not the second
[b]The use of these prefixes should be avoided, except for the measurement of areas and volumes and for the nontechnical use of centimeter, as for body and clothing measurements

Principal SI units used in mechanics

Quantity	Unit	Symbol	Formula
Acceleration	Meter per second squared	. . .	m/s^2
Angle	Radian	rad	a
Angular acceleration	Radian per second squared	. . .	rad/s^2
Angular velocity	Radian per second	. . .	rad/s
Area	Square meter	. . .	m^2
Density	Kilogram per cubic meter	. . .	kg/m^3
Energy	Joule	J	$N \cdot m$
Force	Newton	N	$kg \cdot m/s^2$
Frequency	Hertz	Hz	s^{-1}
Impulse	Newton-second	. . .	$kg\ m/s$
Length	Meter	m	b
Mass	Kilogram	kg	b
Moment of a force	Newton-meter	. . .	$N\ m$
Power	Watt	W	J/s
Pressure	Pascal	Pa	N/m^2
Stress	Pascal	Pa	N/m^2
Time	Second	s	b
Velocity	Meter per second	. . .	m/s
Volume, solids	Cubic meter	. . .	m^3
Liquids	Liter	L	$10^{-3}\ m^3$
Work	Joule	J	$N\ m$

[a]Supplementary unit (1 revolution $= 2\pi$ rad $= 360°$)
[b]Base unit

U.S. customary units and their SI equivalents

Quantity	U.S. customary unit	SI equivalent
Acceleration	ft/s^2	$0.3048\ m/s^2$
	$in./s^2$	$0.0254\ m/s^2$
Area	ft^2	$0.0929\ m^2$
	$in.^2$	$645.2\ mm^2$
Energy	$ft \times lb$	1.356 J
Force	Kip	4.448 kN
	Lb	4.448 N
	Oz	0.2780 N
Impulse	$lb \times s$	$4.448\ N \times s$
Length	Ft	0.3048 m
	in.	25.40 mm
	Mi	1.609 km
Mass	oz mass	28.35 g
	lb mass	0.4536 kg
	Slug	14.59 kg
	Ton	907.2 kg

(continued)

(continued)

Quantity	U.S. customary unit	SI equivalent
Moment of a force	$lb \times ft$	$1.356 \ N \times m$
	$lb \times in.$	$0.1130 \ N \times m$
Moment of inertia		
Of an area	$in.^4$	$0.4162 \times 10^6 \ mm^4$
Of a mass	$lb \times ft \times s^2$	$1.356 \ kg \times m^2$
Momentum	$lb \times s$	$4.448 \ kg \times m/s$
Power	$ft \times lb/s$	$1.356 \ W$
	Hp	$745.7 \ W$
Pressure or stress	lb/ft^2	$47.88 \ Pa$
	$lb/in.^2$ (psi)	$6.895 \ kPa$
Velocity	ft/s	$0.3048 \ m/s$
	$in./s$	$0.0254 \ m/s$
	mi/h (mph)	$0.4470 \ m/s$
	mi/h (mph)	$1.609 \ km/h$
Volume	ft^3	$0.02832 \ m^3$
	$in.^3$	$16.39 \ cm^3$
Liquids	gal	$3.785 \ L$
	qt	$0.9464 \ L$
Work	$ft \times lb$	$1.356 \ J$

Force Systems on Structures

2

Overview

The task in designing any type of a structure requires a deep understanding of forces acting on the parts of the structure. For this purpose the designer must study the types of external forces and take into account the principles of statics. Statics is mostly concerned with the reaction of the body under the influence of external forces. Once we understand the nature and effects of forces on the body, then we will be able to study the material properties of an entire structure under the forces acting on it and show a proper design for the application of interest.

Learning Objectives

Upon completion of this chapter, you will be able to define different types of forces (collinear and concurrent) and compute the respective resultants for these types of force systems by the parallelogram and force triangle methods. You will also be able to define and compute equilibrium conditions for concurrent force systems and the concept of a free body diagram. Your knowledge, application, and problem solving skills will be determined by your performance on the chapter test.

Upon completion of this chapter, you will be able to:

- *Define magnitude, direction, line of action, and point of application of force*
- *Define collinear forces*
- *Compute resultant of collinear forces*
- *Define concurrent forces*
- *Compute resultant of concurrent forces*
- *Define force parallelogram and its application*
- *Define force triangle*
- *Define and compute components of a force*
- *Define and compute the equilibrium of a concurrent force system*
- *Define free body and free body diagram*

© Springer International Publishing Switzerland 2015
P. Ghavami, *Mechanics of Materials*, DOI 10.1007/978-3-319-07572-3_2

2.1   Collinear Forces

The point of the body where the force is applied is called the point of application of the force, and the line along which it acts is called its line of action. The sense of the force, that is, the direction in which it acts, is shown by an arrow. When two forces F_1 and F_2 act along the same line, they are called collinear forces (Fig. 2.1).

Fig. 2.1

2.2   Resultant of Collinear Forces

If forces F_1 and F_2 also have the same direction, the magnitude of their resultant R, can be obtained by adding the magnitudes of the two forces.

$$R = F_1 + F_2$$

For example, if $F_1 = 200$ N and $F_2 = 350$ N, then

$$R = 200 + 350 = 550\,\text{N}$$

As a general rule, when a force acts to the right, it is considered positive, and when it acts to the left, it is negative. The rule of determining the resultant of collinear forces may be stated as follows:

The resultant of any number of forces acting along the same straight line is their algebraic sum.

Example 2.1 Find the resultant of the collinear force system shown (Fig. 2.2).

Fig. 2.2

Solution
$$R = +15 + 25 - 20 - 10 = +10\,\text{lb}$$

Thus, the four given forces may be replaced by a single force of 10 lb acting to the right (positive direction).

Example 2.2 Find the resultant of the collinear forces shown (Fig. 2.3).

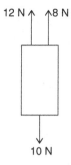

Fig. 2.3

Solution

$$R = +8 + 12 - 10 = 10\,\mathrm{N}$$

Therefore, three given forces may be replaced by a single force of 10 N in the positive direction (upward). Notice that the resultant of two collinear forces, for horizontal or vertical directions, can be found by the graphical method discussed earlier.

Example 2.3 Find the force F that will produce equilibrium for the collinear forces shown (Fig. 2.4).

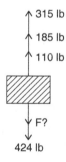

Fig. 2.4

Solution To establish the equilibrium condition, the sum of the forces on one side must be equal to the sum of the forces on the other side, i.e., the resultant of the force system is equal to zero.

$$185 + 110 + 315 - 424 - F = 0$$

Solving algebraically we get,

$$F = 186\,\mathrm{lb}$$

Practice Problems

1. Find the force F that will produce equilibrium for the system of collinear forces shown.

Fig. 2.5

2.

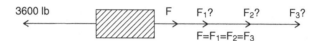

Fig. 2.6

3. Find the resultant of the force systems shown (Figs. 2.7, 2.8, and 2.9) in the x and y directions.

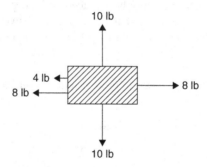

Fig. 2.7

4.

Fig. 2.8

5.

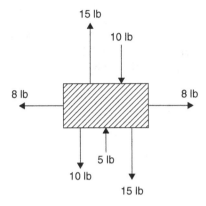

Fig. 2.9

2.3   Concurrent Forces

Concurrent force systems consist of forces whose lines of action pass through a common point (Fig. 2.10).

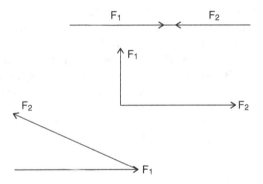

Fig. 2.10

2.4   Resultant of Concurrent Forces

A resultant is a single force which can replace two or more concurrent forces and produce the same effect on the body as the concurrent forces. The resultant will always act at the point of intersection.

The magnitude and direction of the resultant of concurrent force systems can be found by trigonometric or graphical techniques. In the trigonometric method, the law of sines, the law of cosines, and the Pythagorean theorem can be applied. Or it can be determined graphically using a parallelogram, polygon laws, or the force triangle law.

Example 2.4 Two concurrent forces of 30 lb and 40 lb act on a body and make an angle of 90° with each other as shown (Fig. 2.11). Find the magnitude and direction of the resultant (a) graphically and (b) mathematically.

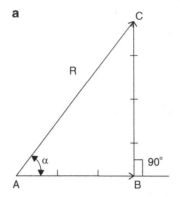
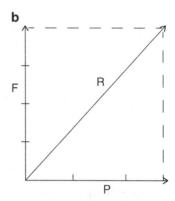

Fig. 2.11

Solution

(a) Graphically: the forces must be drawn to scale. From point A, draw AB equal to 30 lb to any convenient scale and parallel to force P, as shown in Fig. 2.11a, b. From B, draw a line BC in the direction of force F and of a length to represent 40 lb. Then the line AC is the resultant R in magnitude and direction. By measurement, R is found to be 50 lb. Angle α can be measured with a protractor. Figure 2.11a shows the graphical solution by means of force triangle.

(b) Mathematically: using the Pythagorean theorem for the right triangle constructed of P, F, and R, we have

$$
\begin{aligned}
R^2 &= P^2 + F^2 \\
R &= \sqrt{P^2 + F^2} = \sqrt{900^2 + 1,600^2} \\
R &= 50\,\text{lb}
\end{aligned}
$$

The angle α that R makes with P may be found from

$$\tan \alpha = \text{BC}/\text{AB} = F/P = 40/30 = 1.33$$

$$\alpha = 53.13°$$

Example 2.5 Determine the magnitude and direction of the resultant of two concurrent forces $F_1 = 50$ N and $F_2 = 75$ N acting on a body at an angle of 50° with each other (Fig. 2.12). Use the algebraic method only.

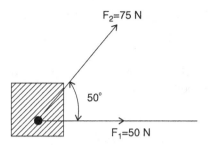

Fig. 2.12

Solution Triangle ABC may be solved for R by using the law of cosines. Find the angle of θ from the geometry (Fig. 2.13).

$$\Theta = 180° - 50° = 130°$$

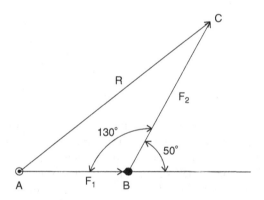

Fig. 2.13

$$\cos 130° = -\cos 50° = -0.643$$

$$
\begin{aligned}
R^2 &= 50^2 + 75^2 - 2(50)(75)(-0.643) \\
&= 2{,}500 + 5{,}625 + 4{,}821 = 12{,}946 \\
R &= 113.8\,\text{N}
\end{aligned}
$$

Angle α may be calculated by the law of sines.

$$F_2/\sin \alpha = R/\sin \theta$$
$$\sin \alpha = (F_2/R)\sin \theta$$

But, $\sin \theta = \sin 130° = \sin 50° = 0.7660$

By substituting in the second equation, we get

$$\alpha = 30.3°$$

Example 2.6 Determine the resultant of the concurrent forces shown (Fig. 2.14) using the method of components.

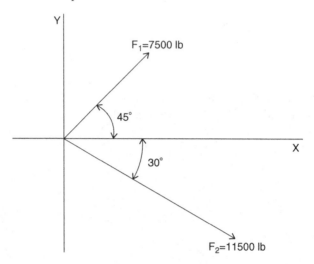

Fig. 2.14

Solution Any force may be resolved into two components at right angles to each other. The components of forces in the x and y directions are determined using trigonometric functions. The forces or component forces in the x and y directions, then, are summed into one single force having $\sum F_x$ and $\sum F_y$ direction. This forms a right triangle with legs equal to $\sum F_x$ and $\sum F_y$ whose hypotenuse is the resultant force in magnitude and direction of the given system of forces. That is,

$$F_x = F \cos \alpha, \quad F_y = F \sin \alpha$$

$$R = \sqrt{\left(\sum F_x\right)^2 + \left(\sum F_y\right)^2}$$

$$\tan \alpha = \Sigma_y / \Sigma_x$$

Therefore,

$$\sum F_x = F_{1x} + F_{2x} \quad = 7,500 \cos 45° + 11,500 \cos 30°$$
$$= 15,262.6 \,\text{lb}$$

$$\sum F_y = F_{1y} + F_{2y} \;\;= 7,500 \sin 45° - 11,500 \sin 30°$$
$$= -446.7\,\text{lb}$$

$$R = \sqrt{\left(\sum F_x\right)^2 + \left(\sum F_y\right)^2} = \sqrt{(15,262.6)^2 + (-446.7)^2}$$

$$R = 15,269 \text{ lb}$$

$$\tan \alpha = \Sigma_y/\Sigma_x = -446.7/15,262.6 = -0.0293$$

where $\alpha \approx 2°$ with respect to the x axis (should take absolute value)

Example 2.7 Determine the magnitude and direction of the resultant of the three concurrent forces shown (Fig. 2.15) using the method of components.

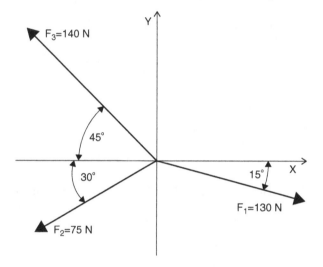

Fig. 2.15

Solution

$$\sum F_x = F_{1x} + F_{2x} + F_{3x} = 130 \cos 15° - 75 \cos 30° - 140 \cos 45°$$
$$= -38.4\,\text{N}$$

$$\sum F_y = F_{1y} + F_{2y} + F_{3y} = -130 \sin 15° - 75 \sin 30° + 140 \sin 45°$$
$$= 27.8\,\text{N}$$

$$R = \sqrt{\left(\sum F_x\right)^2 + \left(\sum F_y\right)^2} = \sqrt{(-38.4)^2 + (27.8)^2}$$
$$R = 47.4\,\text{N}$$

$$\tan \alpha \;= (\Sigma F_y)/(\Sigma F_x) = 27.8/-38.4 = -0.71 \text{ (take absolute value)}$$
$$\alpha \quad = 35.6° \text{ (respect to } x \text{ axis)}$$

2.5   Components of a Force

The resultant of the forces was previously defined as a single force that will produce
the same effect as two forces. Also, it was stated that two forces with their resultant
form a triangle. The converse of this statement is also true: a force may be replaced
by any two forces which, with the given force, form a triangle.

Problems in mechanics of materials are often simplified by resolving forces into
components in the x and y axes and perpendicular to each other. Figure 2.16 shows a
force F. This force can be resolved into two components F_1 and F_2 perpendicular to
each other, where their magnitudes can be calculated knowing the magnitude and
direction of the original vector F.

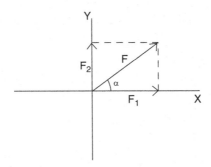

Fig. 2.16

If the angle between F and F_1 is α, then by the right triangle relationship,

$$F_1 = F \cos \alpha$$

and

$$F_2 = F \sin \alpha$$

The components F_1 and F_2 can also be found graphically if they (including force
F) are measured to a proper scale.

Example 2.8 Find the two components of the force F shown (Fig. 2.17) in the
x and y axes, given $F = 25.8$ N. and $\alpha = 35°$.

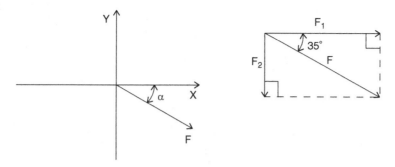

Fig. 2.17

Solution

$$F_1 = F \cos \alpha = (25.8\,\text{N})(\cos 35°) = 21\,\text{N}$$

$$F_2 = -F \sin \alpha = -(25.8\,\text{N})(\sin 35°) = -14.8\,\text{N}$$

Practice Problems

1. Determine the horizontal and vertical components (F_x and F_y) of the following forces.

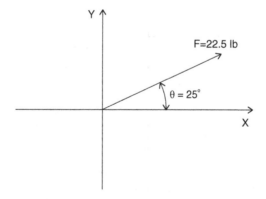

Fig. 2.18

2.

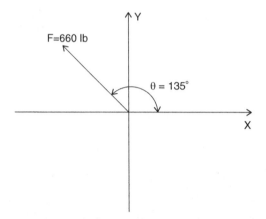

Fig. 2.19

3.

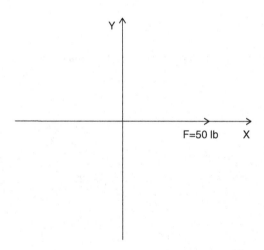

Fig. 2.20

4.

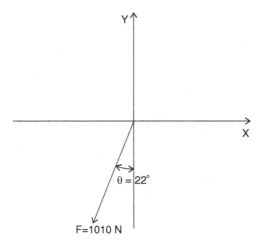

Fig. 2.21

5.

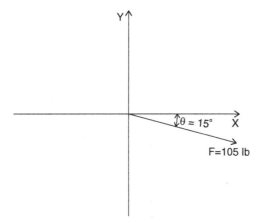

Fig. 2.22

6.

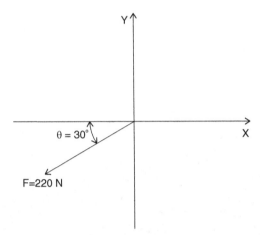

Fig. 2.23

7.

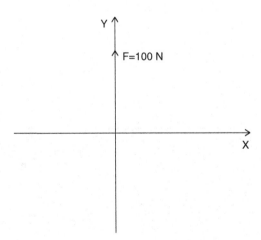

Fig. 2.24

8.

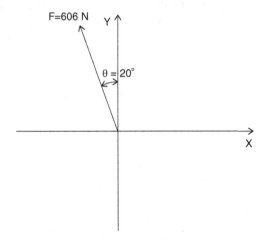

Fig. 2.25

9. A 350 lb wooden crate rests on an inclined plane, making an angle of 30° with the horizontal. Find the magnitudes of the normal force N and force F. These two forces are, respectively, perpendicular to and parallel to the inclined plane (Fig. 2.26).

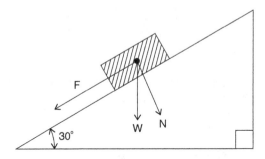

Fig. 2.26

10. Find the magnitude and direction of the resultant of the forces F_1 and F_2 on the beam shown (Fig. 2.27).

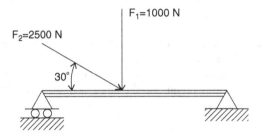

Fig. 2.27

11. Figure 2.28 shows parts of the building structure loaded at each floor level with the given forces of $F_1 = 10,000$ lb, $F_2 = 20,000$ lb, and $F_3 = 25,000$ lb. Find the total vertical load acting on the building foundation.

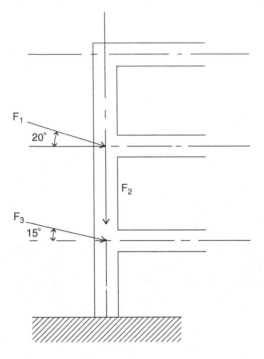

Fig. 2.28

12. Determine the resultant of forces $F_1 = 800$ N at $45°$, $F_2 = 1,200$ N at $60°$, and $F_3 = 1,600$ N at $30°$ by the method of components.

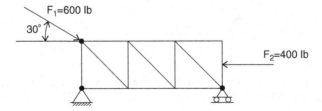

Fig. 2.29

13. Determine the resultant of two forces shown (Fig. 2.29), using the parallelogram law and force triangle.
14. Two forces of 350 and 150 lb make an angle of $120°$ with each other as shown (Fig. 2.30). Find their resultant and the angle it makes with 350 lb force.

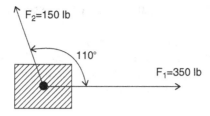

Fig. 2.30

15. Three concurrent forces of 50, 80, and 110 N shown (Fig. 2.31) are in equilibrium condition. Find the angles that these forces make with each other to keep the system in equilibrium.

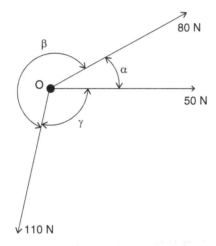

Fig. 2.31

16. Three current forces shown (Fig. 2.32) act on the bolt. Determine the magnitude of the resultant of the forces.

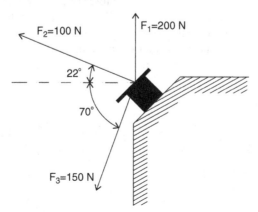

Fig. 2.32

2.6   Equilibrium of Concurrent Force Systems

Equilibrium is a state of balance between opposing forces or actions such that moving a body in one direction is canceled out by other forces that move the body in other directions. Or, in mechanics of materials, a body under the action of a concurrent force system is in equilibrium if the resultant of the force system is equal to zero. In equation form:

$$\sum F_x = 0 \quad \sum F_y = 0$$

When a system of forces is in an equilibrium state, the force polygon of the forces in tip-to-tail fashion will be closed (Fig. 2.33).

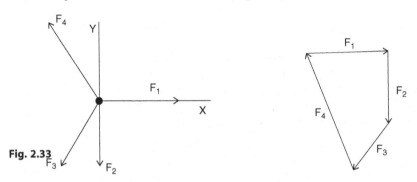

Fig. 2.33

2.7   Equilibrant and the Force Triangle

If two forces are acting on a body, the third force that will hold them in equilibrium is called the equilibrant, or the balancing force. As an example, the relations among three forces in equilibrium could be shown by forces $F_1 = F_2 = 6,000$ lb (Fig. 2.34). These forces are concurrent forces since their lines of action meet at point M. Then, what is the direction and magnitude of a horizontal force F_3 exerted at the same point M such that F_1, F_2, and F_3 will all be in equilibrium?

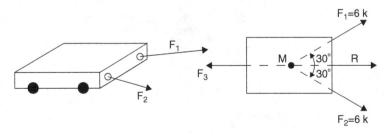

Fig. 2.34

Force F_3, called the equilibrant (or balancing force) of the two forces, is the closing line of the force triangle MNP shown in Fig. 2.35, with the arrowhead pointing in such a direction that the arrowheads of all the forces appear to follow each other around the triangle.

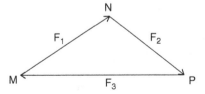

Fig. 2.35

2.8   Mathematical Statement of Equilibrium

In a concurrent force system, equilibrium is met when the summation of the vertical and horizontal components of all forces are zero.

In equation form:

$$R_x = \sum F_x = 0$$

$$R_y = \sum F_y = 0$$

Example 2.9 Find the magnitude and direction of the forces F_1 and F_2 needed in order to produce equilibrium in the diagram shown (Fig. 2.36).

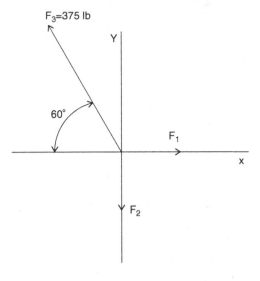

Fig. 2.36

Solution

x-components	y-components
$F_{1x} = F_1$	$F_{1y} = 0$
$F_{2x} = 0$	$F_{2y} = -F_2$
$F_{3x} = -(375\,\text{lb})\cos 60°$	$F_{3y} = (375\,\text{lb})\sin 60°$
$F_{3x} = -188\,\text{lb}$	$F_{3y} = 325\,\text{lb}$

Equilibrium conditions:

$$\sum F_x = 0$$

$$F_1 + 0 + (-188\,\text{lb}) = 0$$

and $F_1 = 188\,\text{lb}$

$$\sum F_y = 0$$

$$0 + (-F_2) + 325\,\text{lb} + 0$$

and $F_2 = 325\,\text{lb}$

Example 2.10 Two cables are connected together at point C and loaded as shown (Fig. 2.37). Determine the tensions in AC and BC. Use the mathematical method and the idea of a force triangle.

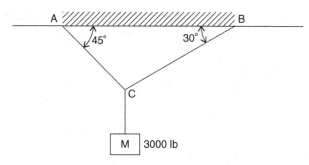

Fig. 2.37

Solution

$$\sum F_x = 0$$

$$\text{BC}\cos 30° - \text{AC}\cos 45° + 0 = 0, \quad \text{or} \quad 0.866\,\text{BC} - 0.707\,\text{AC} = 0 \qquad (2.1)$$

$$\sum F_y = 0$$

$$BC \sin 30 + AC \sin 45 - 3,000 = 0, \quad \text{or} \quad 0.5\,BC + 0.707\,AC = 3,000 \quad (2.2)$$

Equations (2.1) and (2.2) can be solved simultaneously.

$$AC = 2,690\,\text{lb}$$

$$BC = 2,196\,\text{lb}$$

Using the force triangle method, we draw three concurrent vectors tip to tail until the triangle is closed. In this case the resultant of the forces is zero. AC and BC can be obtained using the law of sines.

$$3,000/\sin 75 = AC/\sin 60$$

Solving for AC, we get

$$AC = 2,690\,\text{lb}$$

and

$$3,000/\sin 75 = BC/\sin 45$$

$$BC = 2,196\,\text{lb}$$

2.9 Action and Reaction

According to Newton's third law, if a particle exerts a force on another particle, then the second particle exerts a collinear force of equal magnitude and opposite direction on the first particle. The wall resists the push with an equal and opposite force, too; this is called a reaction. A body weighing 10 lb rests on a table. The action force is the downward pull due to the earth's attraction. The table exerts an upward reaction N and $N = 10$ lb, as the forces are in equilibrium (Fig. 2.38).

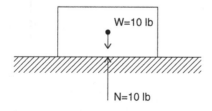

Fig. 2.38

2.10   Newton's First Law and Free Body Diagram

Newton's First Law: If a particle is acted on by forces whose resultant is zero, it remains at rest or will move with constant speed along a straight line. This idea can be utilized in analysis of actual physical engineering problems. These problems can simply be converted into an equilibrium system of forces acting on the particle. In this determination, a free body diagram will be very helpful. A free body diagram is a diagram that represents all the forces acting on a particle.

As an example, consider a block W hanging from two cables tied together at C (Fig. 2.39a). To determine the tension in AC and BC, we construct the free body diagram of the forces shown in Fig. 2.39b.

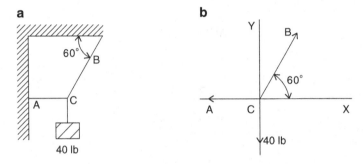

Fig. 2.39

Example 2.11 A 300-N weight hangs from a cord tied to two other cables as shown in Fig. 2.40. Determine the tension in cables A and B.

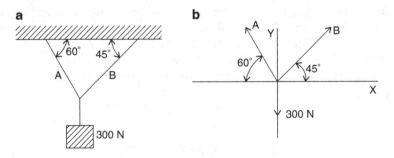

Fig. 2.40

Solution

x-components	y-components
$A_x = -A\cos 60°$	$A_y = A\sin 60°$
$B_x = B\cos 45°$	$B_y = B\sin 45°$
$C_x = 0$	$C_y = -300\,\text{N}$

$$\sum F_x = 0$$

$$-A \cos 60° + B \cos 45° = 0$$

$$\sum F_y = 0$$

$$A \sin 60° + B \sin 45° - 300 = 0$$

Substituting the values in Eqs. (2.3) and (2.4), we have:

$$-0.5A + 0.707B = 0 \qquad (2.3)$$

$$0.866A + 0.707B - 300 = 0 \qquad (2.4)$$

Solving Eqs. (2.3) and (2.4) simultaneously, we get:

$$A = 220\,\text{N}$$

$$B = 155\,\text{N}$$

Example 2.12 Determine the tension in cable AB and the compression in the strut BC shown in Fig. 2.41.

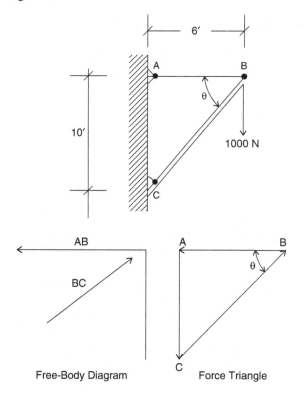

Fig. 2.41

Solution Draw the free body diagram that includes the tension in the cable, compression in the strut, and a 1,000-N weight.

Equations of equilibrium:

$$\sum F_x = 0$$

$$-AB \cos 0 + BC \cos 59° = 0 \tag{2.5}$$

$$\sum F_y = 0$$

$$BC \sin 59° - 1,000 = 0 \tag{2.6}$$

From Eq. (2.6) we find:

$$BC = 1,166.6\,N$$

Substituting for BC into Eq. (2.5) we get

$$-AB + 1,166.6 \cos 59° = 0 \quad \text{and} \quad AB = 600.9\,N$$

Practice Problems

1. Find the tension in each cable shown (Fig. 2.42).

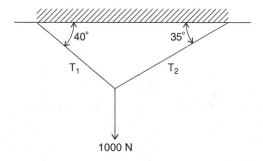

Fig. 2.42

2. For Figs. 2.43 and 2.44, close the force polygon and show the equilibrant as a balancing force.

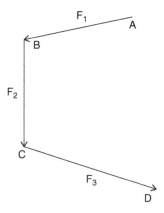

Fig. 2.43

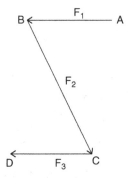

Fig. 2.44

3. Determine the tension in the cable and the compression in the support shown (Fig. 2.45).

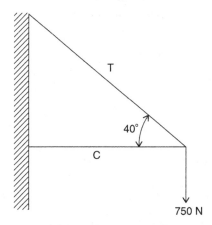

Fig. 2.45

4. A 50-kg sign is suspended by two cables as shown in Fig. 2.46. Find the tension in each cable.

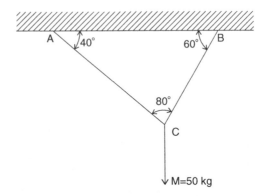

Fig. 2.46

Chapter Summary

1. The forces may apply in the same line of action; these are called collinear forces.
2. The resultant of collinear forces is obtained by adding their magnitudes. The resultant can be obtained algebraically or graphically.
3. The definition and concept of coplanar concurrent forces.
4. The resultant of concurrent forces can be computed using algebraic or graphical methods.
5. The definition of components of a force, and how to find them on the x–y plane.
6. The definition the equilibrium of concurrent forces and equilibrium conditions for these force systems.
7. The free body diagram of the body under influence of force systems.
8. The procedure for how to draw a free body diagram.
9. When a force system is in equilibrium, the force polygon must close.

Review Questions

1. What are *collinear forces*?
2. What are *coplanar concurrent forces*?
3. What are the *components of a force*?
4. What is equilibrium?
5. What is an *equilibrant*?
6. What is a *free body diagram*?
7. What is the principle of *action–reaction in a force system*?
8. What is a *force polygon*?
9. What is the force triangle in the equilibrium problems?
10. What is the Pythagorean theorem and its application?

Problems

1. Determine the resultant of the concurrent force systems shown below.

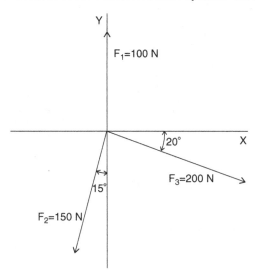

Fig. 2.47

2.

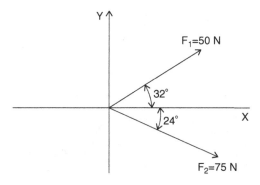

Fig. 2.48

3.

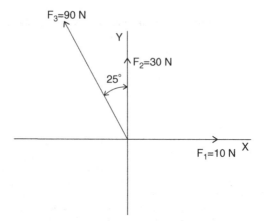

Fig. 2.49

4.

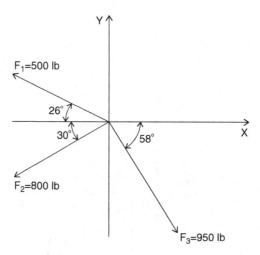

Fig. 2.50

5.

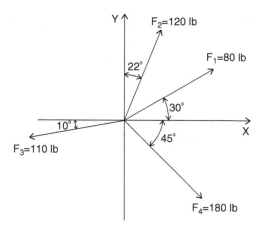

Fig. 2.51

6.

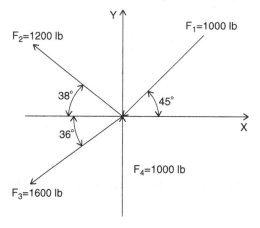

Fig. 2.52

7.

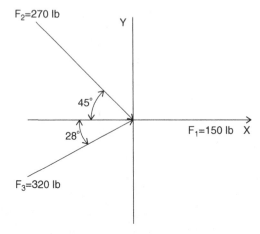

Fig. 2.53

8. Using the law of cosines and the law of sines, find the resultant of the given force triangle (Fig. 2.54).

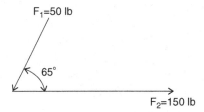

Fig. 2.54

9. Using the method of components, find the resultant of the force system shown (Fig. 2.55).

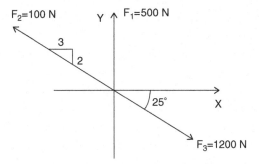

Fig. 2.55

10. Draw a force polygon diagram to show the resultant of the forces shown (Fig. 2.56).

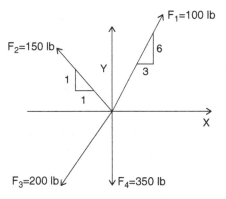

Fig. 2.56

11. A weight of 800 N is tied to the cables below as shown (Fig. 2.57). Find the tension force in the cables.

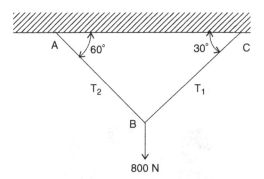

Fig. 2.57

12. For the structure shown below, find the tension on AC and BC (Fig. 2.58).

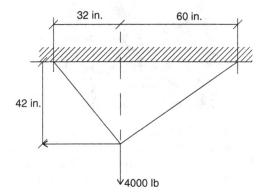

Fig. 2.58

13. Determine the compression in the strut and the tension in the cable shown, assuming the weight is 1,000 N (Fig. 2.59).

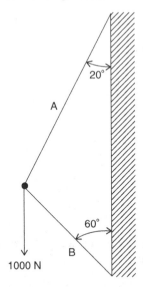

Fig. 2.59

Isaac Newton (1642–1727)

Sir Isaac Newton was born on Christmas Day of 1642 in the hamlet of Woolsthorpe, England, where his mother managed the farm left by her husband who died two months before Isaac was born. The baby was frail and sickly, but he somehow managed to survive and grow stronger, even though he never enjoyed excellent health. He was educated at a local school of low educational standards and as a youth showed no special flair except for an interest in mechanical devices. He was a curious boy and an average student in the grammar school at Grantham. Newton did well enough in his subjects to be admitted to Trinity College of Cambridge in 1661.

When Newton received his bachelor's degree in April 1665, he had to leave Cambridge because the university was closed down, and plague was widespread in the London area. He returned to his home at Woolsthorpe, where he spent the next two years contemplating the ideas about space and time and motion he had first considered while at the university.

By the time of his return to Cambridge (in 1667), it is tolerably certain that he had already firmly laid the foundations of his work in the three great fields with which his name is forever associated: calculus, the nature of light, and universal gravitation and its consequences. Newton's theory of gravitation was based on his theory that "the rate of fall was proportional to the strength of the gravitational force and that this force fell off according to the square of the distance from the center of earth."

His observation of an apple falling from a tree to the ground while at Woolsthorpe led Newton to conclude that the earth was pulling on the apple and the apple on the earth. Newton was the first to speculate that the force that caused the apple to fall to the ground is the same force that keeps the moon in its orbit around the earth and the earth in its orbit around the sun. Newton not only unified and completed the mechanics of Galileo and Kepler, but he showed that the dynamic motions of the universe can be described by basic mathematical relationships that are valid anywhere in the universe. Newton's achievements dominated science and philosophy for the next two centuries.

At first Newton did not publish his discoveries. He is described by most of his contemporaries, including Robert Hooke and Huygens, as having had an abnormal fear of criticism. Later, his astronomer friend Edmond Halley (1656–1742) recognized his greatness and encouraged him to publish his results. This caused Newton to begin working on a book explaining his theory of gravitation, as well as the three laws of motion. He finished the manuscript in 18 months, and it was published in 1687 at Halley's expense as the *Philosophiae Naturalis Principia Mathematica* (*The Mathematical Principles of Natural Philosophy*). The book is usually referred to simply as the *Principia*.

Written in the form of a series of densely worked geometrical axioms and proofs, it remains the greatest and most influential scientific work ever written. It offered a vision of a universe wound up by a cosmic hand and left to run down on its own with all dynamic motions governed by the law of gravitation. The *Principia* brought Newton worldwide fame and ensured his unequalled reputation in the scientific community.

The *Principia* is divided into three books. In a prefatory section Newton defines concepts of mechanics such as inertia, momentum, and force; he then states the three famous laws of motion, which are:

Law I A body continues in its state of rest, or of uniform motion in a right line, unless it is compelled to change that state by forces impressed upon it.

Law II The change (in the quantity) of motion is proportional to the motive power impressed and is made in the direction of the right line in which that force is impressed.

In the definition of motion, Newton means the mass times the velocity. Hence the change in motion, if the mass is constant, is the change in velocity, i.e., the acceleration.

Law III For every action, there is always an opposite and equal reaction, or the mutual actions of the two bodies upon each other are always equal and directed to contrary parts. Although his discoveries were among many made during the Scientific Revolution, Isaac Newton's universal principles of gravity found no parallels in science at the time.

Newton's legacy to modern science is rivaled only by the work of Albert Einstein, who in the twentieth century, would overturn Newton's concept of the universe, stating that space, distance, and motion were not absolute but relative, and that the universe was more fantastic than Newton had ever conceived.

Moment of a Force

3

Overview

A complete study of force systems and their applications requires understanding of the fundamental concept of moment. We will define the moment of a force and how to calculate it. The moment of a force is the tendency of the force to produce rotation about any axis.

Learning Objectives

Upon completion of this chapter, you will be able to define and compute moment of a force for many systems and to apply the principle of moments to find the resultant of a non-concurrent force system. You will also be able to define the equilibrium of parallel forces in a plane, and compute support reactions in beams with the application of moment of forces. Your knowledge, application, and problem solving skills will be determined by your performance on the chapter test.

Upon completion of this chapter, you will be able to:

- *Define and compute moment of a force with respect to a point*
- *Identify positive moment or negative moment*
- *Define the principle of moments*
- *Identify the unit of moment in different systems of units*
- *Identify equilibrium of parallel forces in a plane*
- *Compute support reactions in beams using the application of moment of forces*

3.1   Moment of a Force

Moment of a force F about a given point O is the product of the force F and its perpendicular distance r from the line of action of the force to the center of rotation O.

© Springer International Publishing Switzerland 2015

P. Ghavami, *Mechanics of Materials*, DOI 10.1007/978-3-319-07572-3_3

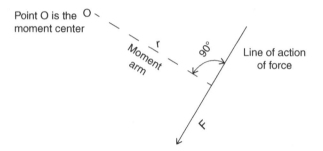

Point O is the
moment center

Fig. 3.1

Moment of force = magnitude of force × moment arm

or

$$M = F \times r,$$

where

$F = $ force
$r = $ moment arm

Unit of moment in the English system is lb-ft and in the metric system is N.m.

3.2   Sign Convention of a Moment

In the figure shown (Fig. 3.2), it is evident that the 100-lb force tends to rotate about point A in one direction, and about point B in the opposite direction. Some rule for the direction of rotation is then necessary.

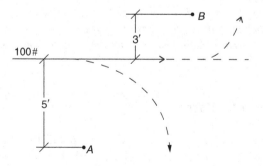

Fig. 3.2

Clockwise rotation (+) Counterclockwise rotation (−)

$$\text{Moment of } 100\,\text{lb. force about point A} \; = (+100\,\text{lb})\,(5\,\text{ft})$$
$$= +500\,\text{lb-ft}$$

$$\text{Moment of } 100\,\text{lb. force about point B} \; = (-100\,\text{lb})\,(3\,\text{ft})$$
$$= -300\,\text{lb-ft}$$

Example 3.1 Find the moment of forces $F_1 = 500$ N and $F_2 = 750$ N with respect to point O. Assume $r_1 = 2$ m and $r_2 = 3$ m (Fig. 3.3).

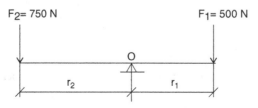

Fig. 3.3

Solution

$$\text{Moment of force } F_1 : \quad M_1 = (500\,\text{N}\,)(2\,\text{m}) = 1,000\,\text{N.m}$$

$$\text{Moment of force } F_2 : \quad M_2 = (-750\,\text{N})(3\,\text{m}) = -2,250\,\text{N.m}$$

Example 3.2 Find the moment of forces shown in Fig. 3.4 with respect to point O.

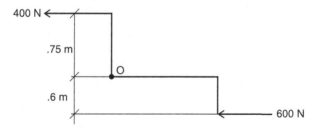

Fig. 3.4

Solution A force of 600 N produces a positive moment about point O, and a force of 400 N produces a negative moment about point O.

$$M_o = (600\,\text{N}) \times (0.6\,\text{m}) = 360\,\text{N.m}$$

$$M_o = (400\,\text{N}) \times (0.75\,\text{m}) = -300\,\text{N.m}$$

Example 3.3 Find the moment of force for $F = 250$ lb shown (Fig. 3.5) with respect to point A and point B.

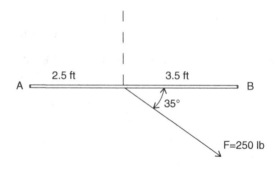

Fig. 3.5

Solution As mentioned earlier, the moment of a force must be perpendicular to the moment arm.

Now, we can resolve the force into its x and y components (F_x and F_y) as shown in Fig. 3.6. F_x, however, does not contribute to the moment of force F, because it is not perpendicular to the moment arm. In this case, only F_y will contribute to the calculation of moment of force F.

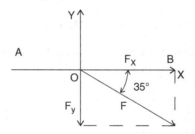

Fig. 3.6

$$M_a = (250\,\text{lb})\,\sin 35°(2.5\,\text{ft}) = +358.5\,\text{lb-ft}$$

$$M_b = (250\,\text{lb})\,\sin 35°(3.5\,\text{ft}) = -501.9\,\text{lb-ft}$$

Example 3.4 A force of 120 N is applied to a pulley ($D = 45$ cm) as shown in Fig. 3.7. What is the moment of force about the center point O?

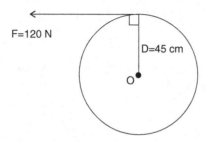

Fig. 3.7

Solution The moment of force, in this case, is negative because it is in the counterclockwise direction.

$$M_o = -(120\,\text{N})(0.45\,\text{m}) = -54\,\text{N.m}$$

3.3 Principle of Moments

The principle of moments, called Varignon's theorem, states that the moment of a force about a point is exactly equal to the sum of the moments of the components of that force about the point. This principle is used in calculating the moment of a force when it is too difficult to calculate the perpendicular distance (see Example 3.3). A variety of structural problems in mechanics of materials can be dealt with through the principle of moments by using the rectangular components of the force.

Example 3.5 Determine the moment of the 200-N force shown in Fig. 3.8 with respect to point O.

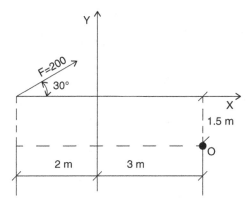

Fig. 3.8

Solution The easiest way to solve the problem is to use the principle of moments. In this case we need to resolve the given 200-N force the along x and y axes. The components are:

$$F_x = 200 \cos 30° = 173\,\text{N}$$

$$F_y = 200 \sin 30° = 100\,\text{N}$$

and the moment of force F, using the principle of moments, is the sum of the moments of force components about the point O. Then we have:

$$M_o = 173 \times 1.5 + 100 \times 5 = 760\,\text{N.m}$$

3.4   Moment Equation for Equilibrium

The moment equation for equilibrium with respect to a point states that the sum of the moments of all the forces about that point is equal to zero. In other words, the resultant moment about that point is zero. That means:

$$\sum M = 0$$

Example 3.6 The condition of moment equilibrium about a pivot point O (Fig. 3.9) of two forces F_1 and F_2 is:

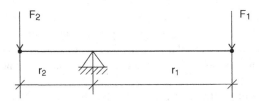

Fig. 3.9

$$F_1 \times r_1 - F_2 \times r_2 = 0 \qquad (3.1)$$

or

$$F_1 \times r_1 = F_2 \times r_2 \qquad (3.2)$$

If
$F_1 = 180$ N
$F_2 = 300$ N
$r_1 = 320$ cm
From Eq. (3.2), we get,

$$r_2 \quad = (F_1 \times r_1)/F_2 = (180\,\mathrm{N} \times 320\,\mathrm{cm})/300\,\mathrm{N}$$
$$= 192\,\mathrm{cm} = 1.92\,\mathrm{m}$$

The principle of moment equilibrium was known thousands years ago, and the lever was used in the construction of the pyramids as well as in a variety of ingenious devices and mechanisms.

Example 3.7 Given a group of parallel forces $F_1 = 2{,}000$ lb, $F_2 = 5{,}000$ lb, $F_3 = 3{,}000$ lb acting on a beam as shown Fig. 3.10, find:

(a) The resultant of the forces.
(b) The distance of the resultant from point O (left end of the beam) (assume $x_1 = 2$ ft, $x_2 = 5$ ft, and $x_3 = 8$ ft)

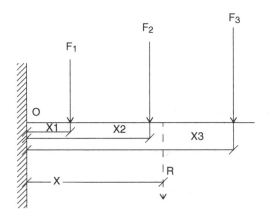

Fig. 3.10

Solution

(a) $R = \sum F_y = F_1 + F_2 + F_3 = 2{,}000\,\text{lb} + 5{,}000\,\text{lb} + 3{,}000\,\text{lb} = 10{,}000\,\text{lb}$
   The resultant R is also acting downward.
(b) Writing the moment equation for equilibrium, we obtain

$$\sum M_o = Rx = 2{,}000(2) + 5{,}000(5) + 3{,}000(8) = 53{,}000\,\text{lb-ft}$$

Substituting $R = 10{,}000$ lb, we have

$$10{,}000\,x = 53{,}000\,\text{lb-ft}$$

and

$$x = 53{,}000\,\text{lb-ft}/10{,}000 = 5.3\,\text{ft}$$

Example 3.8 Figure 3.11 shows a bar AB hanging from the ceiling with two unknown forces F_1 and F_2. Determine the forces if the given weight $W = 600$ N. Assume the system is in equilibrium.

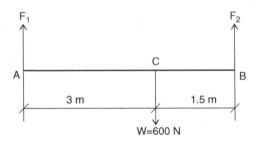

Fig. 3.11

Solution Using the principle of moments for equilibrium for the system of forces with respect to point B, we obtain

$$\sum M_b = 0$$

$$-600(1.5) + F_1(4.5) = 0$$

and

$$F_1 = 600(1.5)/4.5 = 200\,\text{N}$$

If we take the moment of forces with respect to point A using moment equilibrium conditions, we get

$$\sum M_a = 0$$

$$600(3) - F_2(4.5) = 0$$

and

$$F_2 = 600(3)/4.5 = 400\,\text{N}$$

We can also find F_2 knowing that $F_1 + F_2$ equals the load ($W = 600$ N) on the bar.

Practice Problems

1. Determine the moment of given forces with respect to point O in Fig. 3.12.

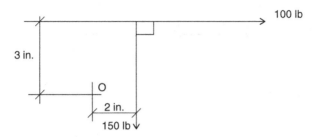

Fig. 3.12

2. Determine the moments of the given force $F = 800$ N with respect to point A and point B (Fig. 3.13).

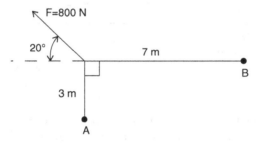

Fig. 3.13

3. For the non-concurrent forces shown in Fig. 3.14, find (a) the moment of forces with respect to point A, and (b) the magnitude and direction of the resultant.

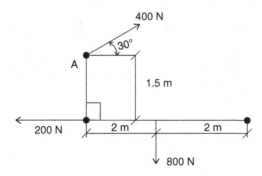

Fig. 3.14

4. Find the forces F_1 and F_2 shown in Fig. 3.15 to produce the equilibrium in the bar system.

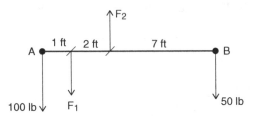

Fig. 3.15

5. Determine the force F necessary to open the safety valve in the figure shown (Fig. 3.16). Assume that the suspended weight exerts a 200-N force on the lever.

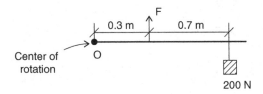

Fig. 3.16

6. Compute the moment of forces with respect to points B and C for the beam shown in Fig. 3.17. Is this system in equilibrium? Show your proof mathematically.

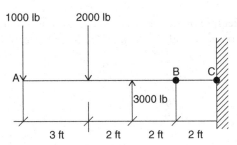

Fig. 3.17

7. A force of 150 N is applied at the end of a 1.5 m lever (Fig. 3.18). Find the moment of force with respect to point B.

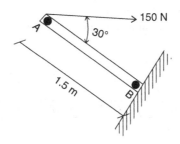

Fig. 3.18

8. Compute the moment of forces shown (Fig. 3.19) with respect to point A and point B. Is this force system in equilibrium? Show your proof mathematically.

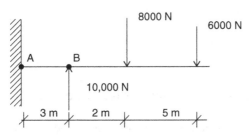

Fig. 3.19

9. Determine (a) the resultant of the forces acting on the structure shown in Fig. 3.20 and (b) the distance of the resultant force to the point A on the footing. Assume $F = 300$ N.

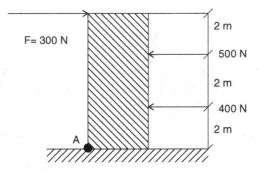

Fig. 3.20

10. Figure 3.21 shows a construction crane with a boom of 15 m and a 10,000-N balancing weight at the other end that is positioned 4 m from the center of the crane. Determine the amount of the construction load to be lifted to bring the system into equilibrium.

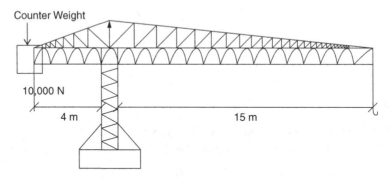

Fig. 3.21

3.5   Equilibrium of Parallel Coplanar Forces

Forces whose lines of actions are not intersecting each other are called parallel forces. The resultant of parallel forces which are in equilibrium is equal to zero. In fact, the equilibrium equations for parallel forces are

$$\sum F_x = 0$$

$$\sum F_y = 0$$

$$\sum M = 0$$

Note that in a parallel force system, the problem cannot be solved without using the moment equation. That means the first two equations above are necessary for equilibrium conditions, but are not themselves sufficient. With the moment equation, the force system equilibrium will be established.

The principle involved in the study and application of parallel forces is best understood by a simple example. Let's look at the following beam (Fig. 3.22) experiencing three forces at points A, B, and C. We can check whether the beam is in equilibrium.

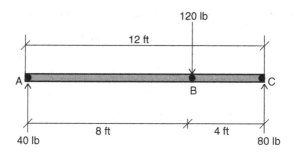

Fig. 3.22

The sum of the forces is: $40 + 80 - 120 = 0$.
We compute the moment of forces about points A, B, and C as follows:

$$\sum M_a = (8 \times 120) - (12 \times 80) = 960 - 960 = 0$$

$$\sum M_b = (8 \times 40) - (4 \times 80) = 320 - 320 = 0$$

$$\sum M_c = (12 \times 40) - (4 \times 120) = 480 - 480 = 0$$

These results illustrate the principle that the above parallel forces are in equilibrium, and that the sum of their moments about any axis through any point is zero. In other words, if a body is in equilibrium, it must neither translate nor rotate. Then the conditions for static equilibrium are as follows:

1. The algebraic sum of the forces must be equal to zero.

$$\sum F = 0$$

2. The algebraic sum of the moments about any axis through any point must be equal to zero.

$$\sum M = 0$$

3.6   Applications of Moment of Forces

One of the applications of the moment of forces is to determine the reactions in a beam-loaded system. The loads can be concentrated or distributed loads. A special type of distributed load is a uniformly distributed load, or a load with constant density. Examples of these types of loading will be shown later.

Example 3.9 Find the reactions of the supports for the beam shown in Fig. 3.23.

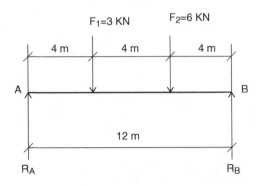

Fig. 3.23

Solution By the conditions of equilibrium, the algebraic sum of the moments about any point is zero. A convenient point to choose would be either the right or left support, since one of the unknowns will be eliminated. If point A is chosen, then

$$\sum M_a = -R_B(12) + 6(8) + 3(4) = 0$$

$$-12R_B = -60$$

and

$$R_B = 5\,\text{KN},$$

but

$$R_A + R_B = 3 + 6 = 9\,\text{KN} \quad (\text{first condition for equilibrium})$$

Solving for R_A, we obtain

$$R_A = 9 - 5$$
$$= 4\,\text{KN}$$

Example 3.10 Compute the magnitude of the reactions for the beam shown in Fig. 3.24. The supports are a pin support at point A and a roller at end B.

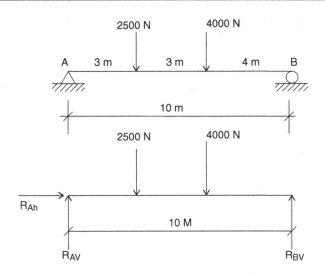

Fig. 3.24

Solution The free diagram at the pin (A) shows that there are two reactions here, R_{Ah} and R_{Av}, to be considered.

$$\sum F_x = 0$$

$$R_{Ah} = 0, \quad \text{there is no horizontal component}$$

$$\sum F_y = 0$$

$$R_{Av} + R_{Bv} - 2,500 - 4,000 = 0$$

Rearranging the equation

$$R_{Av} + R_{Bv} = 6,500$$

$$\sum M_a = 0$$

$$2,500(3) + 4,000(6) - R_{Bv}(10) = 0$$

Solving for R_{Bv}, gives,

$$R_{Bv} = 3,150\,\text{N}$$

Substituting into equation $\sum F_y = 0$, we get

$$R_{Av} = 3,350\,\text{N}$$

Example 3.11 Find the reactions at A and B exerted by the walls as shown in Fig. 3.25. Draw the free body diagram and show the forces including the reactions on the beam.

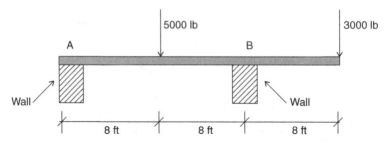

Fig. 3.25

Solution

$$\sum F_y = 0 \quad (\text{1st condition of equilibrium})$$

$$R_A + R_B = 5,000 + 3,000 = 8,000 \text{ lb}.$$

$$\sum M_a = 0$$

$$5,000(8) - R_B(16) + 3,000(24) = 0$$

Solving the equation gives

$$R_B = 7,000 \text{ lb}$$

Substituting into equation $\sum F_y = 0$, we have

$$R_A = 1,000 \text{ lb}$$

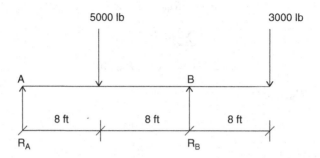

Fig. 3.26 Free body diagram

3.7   Distributed Load Systems

There are many problems in mechanics of materials that deal with certain types of distributed load. Examples of distributed load are things like roof loads, floor loads, wind loads, and snow loads. If the load is distributed uniformly along the length, it is called a uniformly distributed load. Distributed loads can also be non-uniform. Distributed loads normally are depicted by a diagram, called a load diagram. The resultant of a distributed load is simply the area of the load diagram (Fig. 3.27).

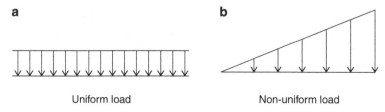

Fig. 3.27

Example 3.12 Compute the total amount of the uniformly distributed load shown in Fig. 3.28.

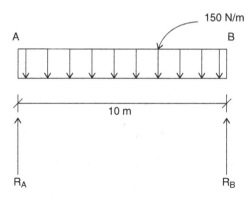

Fig. 3.28

Solution The density of the load is 150 N/m, and the total amount of the distributed load is obtained through multiplying the load density by the meters length of the load distributed on the beam:

$$Total\,load = 150\,N/m \times 10\,m = 1,500\,N$$

Example 3.13 Compute the magnitude of the reactions for the beam shown (Fig. 2.29) resting on the two end supports and carrying a uniformly distributed load of 1,000 lb/ft, and an additional concentrated load of 4,000 lb.

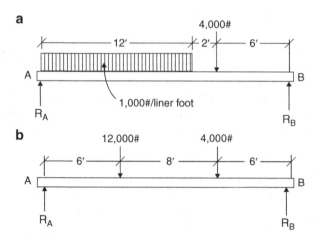

Fig. 3.29

Solution   Take the point B as the center of the moments:

$$20R_A = (1,000 \times 12 \times 14) + (4,000 \times 6)$$

Solving the equation for R_A

$$20R_A = 192,000$$

$$R_A = 9,600\,\text{lb}$$

Using the 1st equation of equilibrium

$$R_A + R_B = 12,000 + 4,000 = 16,000$$

Substituting a value for R_A into above equation, it gives

$$R_B = 16,000 - 9,600 = 6,400\,\text{lb}.$$

Practice Problems

1. A beam resting on two end supports carries concentrated loads as shown in Fig. 3.30. Compute the reactions.

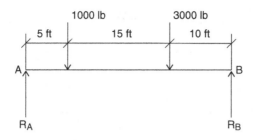

Fig. 3.30

2. Compute the magnitude of reactions for the beams shown in Figs. 3.31 and 3.32.

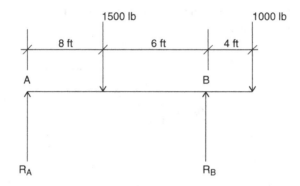

Fig. 3.31

3.

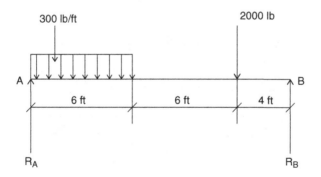

Fig. 3.32

4. A beam 12 ft long is supported at the ends as shown in Fig. 3.33. A load of 600 lb is placed 5 ft from the left end. Compute the reactions.

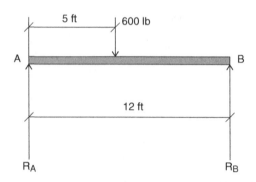

Fig. 3.33

5. What should be the values of the forces R and F for the beam shown in Fig. 3.34 to keep the force system in equilibrium?

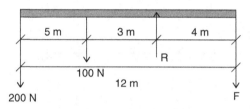

Fig. 3.34

6. Using the principle of moments, determine the values of reactions R_A and R_B for the load shown on the beam in Fig. 3.35.

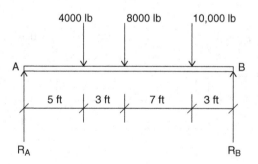

Fig. 3.35

7. Determine the reactions R_A and R_B for the beam shown in Fig. 3.36.

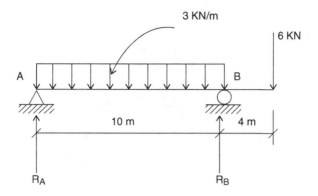

Fig. 3.36

8. Determine the reactions R_A and R_B for the beam shown in Fig. 3.37.

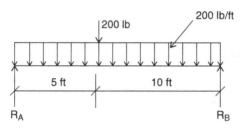

Fig. 3.37

9. Determine the reactions R_A and R_B for the beam shown in Fig. 3.38.

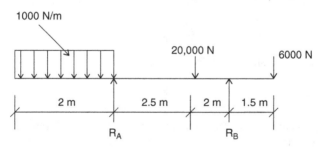

Fig. 3.38

10. Frame A as shown in Fig. 3.39 is loading under a 5,000 N force. Determine the reactions at the A (pin) and B (roller) supports.

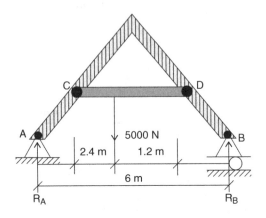

Fig. 3.39

Chapter Summary

1. The moment of a force is a measure of its tendency to rotate a body. It is calculated by the equation:

$$M = F \times r$$

with clockwise direction as positive; and counterclockwise direction as negative.
2. The resultant of a force system is:

$$R = \sum F_i$$

and the resultant moment of a force system is:

$$M = \sum M_i$$

3. If a body is in equilibrium, the sum of the moments of the system of forces about any point is zero:

$$\sum M_i = 0$$

4. For the coplanar parallel force systems, the resultant of forces which is in: equilibrium is equal to zero. Equations of coplanar forces for equilibrium are

$$\sum F_x = 0$$

$$\sum F_y = 0$$

$$\sum M = 0$$

5. The free body diagram is very useful in showing all the forces acting on a body. In fact, it is highly advised to draw a free-body diagram before attempting to solve problems.

Review Questions

1. What is moment of a force?
2. What is the moment arm?
3. What is the moment center?
4. When can the moment of a force be positive, and when can it be negative?
5. What happens to the resultant of coplanar forces on a body when the body is in equilibrium?
6. What is the difference between a coplanar parallel force system and a coplanar non-concurrent force system?
7. What are support reactions of a certain beam?
8. What is the difference between a force and a reaction?
9. How many components of forces must be considered in pin and roller supports?
10. How many types of beam loading are there?

Problems

1. Find the moment of forces shown (Fig. 3.40) with respect to points A and B.

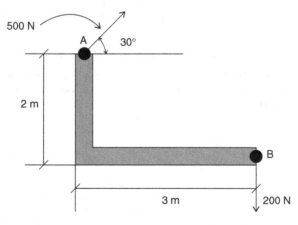

Fig. 3.40

2. A cantilever beam 3 ft long is shown in Fig. 3.41. If a concentrated load of 3,000 lb acts on the free end, calculate the moment of the load with respect to the fixed point A.

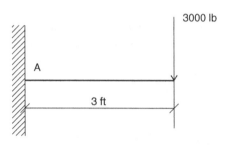

3000 lb

A

3 ft

Fig. 3.41

3. Find the moment force of 20,000 N acting at the top of the tower shown in Fig. 3.42 with respect to points A and B.

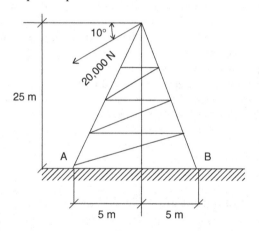

10°

20,000 N

25 m

A   B

5 m   5 m

Fig. 3.42

4. A 30 m high structure with a cross section of 20 m × 20 m is exposed to wind pressure of 5,000 N/m² as shown in Fig. 3.43. Calculate the overturning moment of wind force with respect to point A. Assume the weight of the building is 4×10^6 N.

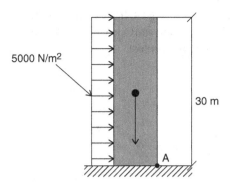

Fig. 3.43

5. Determine the values of reactions R_A and R_B for the force system on the simply supported beam shown in Fig. 3.44.

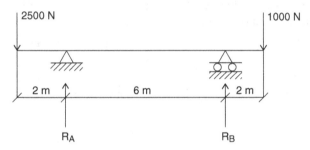

Fig. 3.44

6. Find the horizontal force F necessary to rotate the block about point A as shown in Fig. 3.45.

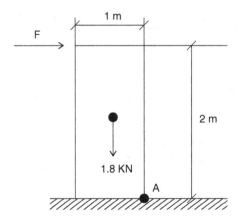

Fig. 3.45

7. A wheel 1 m in diameter (Fig. 3.46) weighs 150 N with its load. Find the
   horizontal force F necessary to start the wheel rolling over an obstruction 30 cm
   high (forces act through the center).

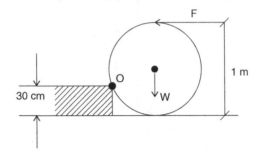

Fig. 3.46

8. In the member shown (Fig. 3.47), calculate the value of force F in Newton(s) to
   keep the member in equilibrium.

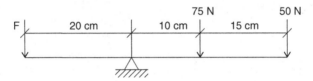

Fig. 3.47

9. Compute the reactions R_A and R_B at the supports A and B of the beams shown
   below (Fig. 3.48).

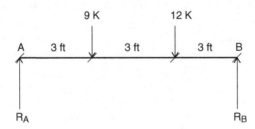

Fig. 3.48

10. Find the sum of the moments of forces $C_x = 7.9$ kips, $C_y = 2.1$ kips, $C = 7.5$ in.
    and $\theta = 28°$ shown (Fig. 3.49) with respect to point O (1 kip = 1,000 lb).

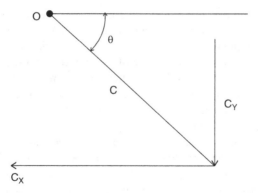

Fig. 3.49

Archimedes of Syracuse (287–212 B.C.)

Of all the Greek philosophers who concerned themselves with physical phenomena, Archimedes of Syracuse was the most notable and was closest to what we now consider a scientist. Archimedes, the son of the astronomer Phidias, was born at Syracuse in 287 B.C., and was good friends with King Hieron, the local ruler.

Although few details of his life are known, he is regarded as one of the leading scientists in classical antiquity. He spent part of his youth in Egypt learning mathematics from the immediate successors of Euclid.

Archimedes combined theory and experiment in a manner similar to scientific procedure today, but no body of basic scientific principles resulted from his work. He was a great experimentalist, a physicist, engineer, inventor, and astronomer.

He tied geometry to mechanics and used geometrical arguments ingeniously to make his proofs. In mechanics he wrote On the Equilibrium of Planes or The Centers of Gravity of Planes, a work in two books. He is most famous for his discovery of the principle of buoyancy (the Archimedes principle), the water snail, Archimedes screw, a helical pump for raising water for irrigation, and the astronomical cross-staff with which he made accurate celestial observations.

He is credited with designing innovative machines, such as compound pulleys, and defensive war machines to protect his native Syracuse from invasion. He demonstrated his mathematical skill by showing how to deduce geometrically the number pi (the ratio of the circumference of a circle to its diameter) to any desired accuracy. He did this by approximating the circumference of a circle with the perimeter of a circumscribed or inscribed many-sided regular polygon.

He was also one of the first to apply mathematics to physical phenomena, founding hydrostatics and statics, including an explanation of the principle of the lever. Unlike Aristotle, whose mechanics is integrated into a theory of physics which goes so far as to incorporate a system of the world, Archimedes made an autonomous theoretical science of statics, based on postulates of experimental origin and afterwards supported by mathematically rigorous demonstrations.

How much Archimedes truly preferred the theoretical to the practical may never be known. It is clear, however, that in his work there is tension between theory and application, a tension that still prevails mathematics 22 centuries later. Archimedes died at the age 75.

Centroid of an Area

4

Overview

For most of the problems in mechanics of materials, it frequently happens that we must determine the position of the centroid; this is accomplished most readily by mathematical methods. This chapter is designed to show what the centroid of an area is and how to find it. Knowing the centroid of an area is very useful in structural design, especially when a beam is subjected to loads that cause it to bend about a neutral axis which passes through the centroid.

Learning Objectives

Upon completion of this chapter, you will be able to define and determine the location of the centroid of a plane figure, and then use the mathematical concept of moment of an area to find the centroid of various shapes. Your knowledge, application, and problem solving skills will be determined by your performance on the chapter test.

Upon completion of this chapter, you will be able to:

- *Define center of gravity*
- *Define and compute the centroid of a simple area*
- *Define and compute the moment of a simple area with respect to an axis*
- *Define a composite area*
- *Compute the moment of a composite area with respect to an axis*
- *Determine the location of the centroid of a composite area*

4.1   Center of Gravity

Imagine that a body is composed of an infinite number of small particles, and each particle has its own gravitational force that acts towards the Earth's center. These forces are assumed to be a parallel force system, and the resultant of these forces (the *weight* of the body) will act through a point, the body's *center of gravity*.

© Springer International Publishing Switzerland 2015
P. Ghavami, *Mechanics of Materials*, DOI 10.1007/978-3-319-07572-3_4

4.2   Centroid of an Area

If we assume that a body is very thin with uniform thickness, then the resultant of forces would be proportional to the area instead of the volume. The location of this resultant is called the *centroid* of the area. Location of the centroid is defined by x and y coordinates from a set a reference axes (Fig. 4.1). The centroid of an area is a very important point in structural engineering design due to the fact that the bending of the objects occurs about the neutral axis where the location of the centroid is.

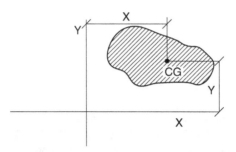

Fig. 4.1

The position of the centroid for symmetrical shapes will be along the line of symmetry. Or, if there are two lines of symmetry, the centroid will be at their point of intersection. We will first show the centroid of a plane area, and then the centroid of a composite area will be discussed later in this chapter.

For example, the centroid of a symmetrical figure such as a rectangle is the point of intersection of the diagonals (Fig. 4.2a), and the centroid of a circular area is its geometrical center, i.e. its center (Fig. 4.2b). The centroid of a triangular area is at a distance equal to one-third of the perpendicular distance measured from any side to the opposite vertex (Fig. 4.2c).

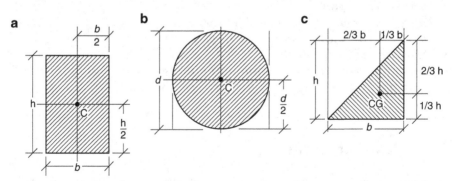

Fig. 4.2

4.3   **Centroid of Simple Areas**

Now we will show you how to find the location of the centroid of some simple geometrical figures. Most of these figures you already know, but some of these figures have formulas whose proofs require calculus, which puts them beyond the scope of this book. However, the student is encouraged to pursue them if he/she is interested.

Example 4.1 Determine the coordinates of the centroid of the triangular area shown in Fig. 4.3.

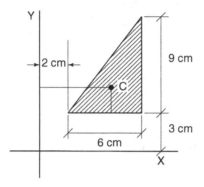

Fig. 4.3

Solution The distance from the centroid to the horizontal side of the triangle is:

$$h/3 = 9/3 = 3$$

and the distance from the centroid to the x axis is:

$$Y_c = 3 + 3 = 6\,\text{cm}$$

The distance from the centroid to the y axis is:

$$X_c = 2/3(6) + 2 = 4\,\text{cm} + 2\,\text{cm} = 6\ \text{cm},$$

so the coordinates of the centroid are C (6, 6).

Example 4.2 Locate the centroids of the areas shown in Fig. 4.4.

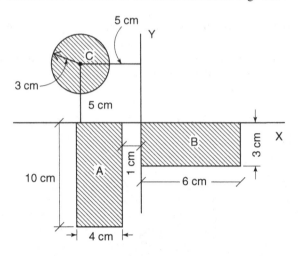

Fig. 4.4

Solution Reading the measurements from Fig. 4.4 for each figure (A, B, C), we get:

Rectangular figure A:

$$X_c = -3 \, cm$$

$$Y_c = -5 \, cm$$

Rectangular figure B:

$$X_c = 3 \, cm$$

$$Y_c = -1.5 \, cm$$

Circular figure C:

$$X_c = -5 \, cm$$

$$Y_c = 5 \, cm$$

Example 4.3 Locate the centroids of the areas shown in Fig. 4.5.

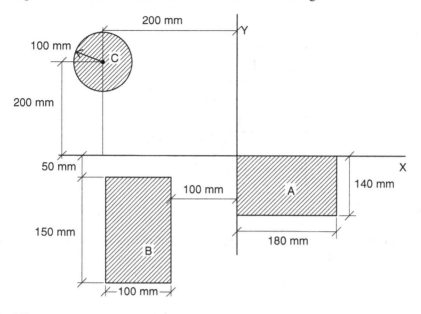

Fig. 4.5

Solution

Rectangular figure A:

$$X_c = 90\,\text{mm}$$

$$Y_c = -70\,\text{mm}$$

Rectangular figure B:

$$X_c = -150\,\text{mm}$$

$$Y_c = -125\,\text{mm}$$

Circular figure C:

$$X_c = -200\,\text{mm}$$

$$Y_c = 200\,\text{mm}$$

Example 4.4 Locate the centroid of the right triangle shown in Fig. 4.6.

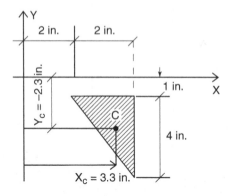

Fig. 4.6

Solution The x and y coordinates of the centroid of the triangle can be found with respect to the x and y axes.

$$X_c = 4 - (0.667) = 3.33 \text{ in.}$$

$$Y_c = -(1 + 1.33) = -2.33 \text{ in.}$$

Centroid of Quadrant

The quadrant of a circle with a radius r has a centroid. The distance from the centroid to the x and y axes is shown below (Fig. 4.7).

$$X_c = 4r/(3\pi)$$

$$Y_c = 4r/(3\pi)$$

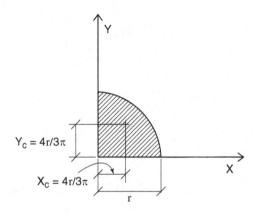

Fig. 4.7

Example 4.5 Locate the centroid of the quadrant shown (Fig. 4.8).

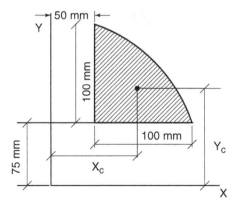

Fig. 4.8

Solution

$$x_c = (100 \times 4/3\pi) + 50\,\text{mm} = 92.4\,\text{mm}$$

$$y_c = (100 \times 4/3\pi) + 75\,\text{mm} = 117.4\,\text{mm}$$

Practice Problem
Locate the centroids of the following simple areas.

1.

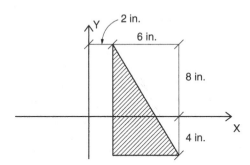

2.

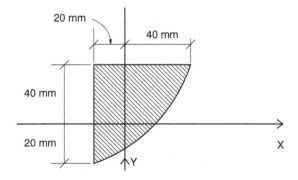

3.

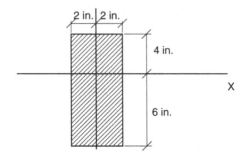

4.

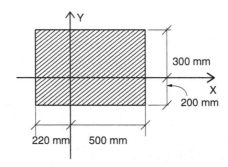

Fig. 4.9

5.

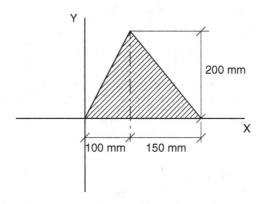

Fig. 4.10

4.4 Moment of a Simple Area

The concept of moment of an area is used to simplify the calculation of the centroid of a plane figure. It is defined as the product of the area and the normal distance of the centroid of the area from a given axis.

The moment of an area can be found with respect to the x axis or the y axis. Suppose we want to calculate the moment of the rectangular area shown (Fig. 4.11) with respect to the x axis. Then we multiply the area by the normal distance from the centroid to the x axis. This mathematically is represented as:

$$M_x = AY_c$$

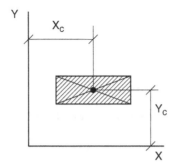

Fig. 4.11

In the same manner, the moment of an area with respect to the y axis is the product of the area and the normal distance from the centroid to the y axis. This is also represented as:

$$M_y = AX_c$$

The unit for the moment of an area in the English system is in.^{2} × in. = in.^{3}, and in the metric system it is mm^2 × mm = mm^3. This kind of dimension may not be conceivable to most of us; it is just a mathematical concept.

Now, to calculate the moment of a simple area, such as a rectangle, triangle, square, or circle, with respect to the x or y axis, we simply multiply the value of the area by the centroid normal distance from the axis to get the moment of an area with respect to the given axis.

Example 4.6 Find the moment of the triangle shown (Fig. 4.12) with respect to the x and y axes.

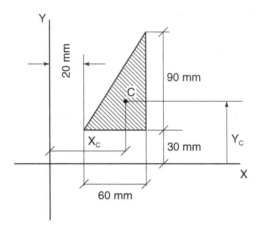

Fig. 4.12

Solution The area of the triangle is

$$A = b \times h \,/2 = (60\,\text{mm} \times 90\,\text{mm})/\,2 \;=\; 2{,}700\,\text{mm}^2$$

The distance from the centroid to the base of the triangle is

$$h/3 = 90/3 = 30\,\text{mm}$$

and the distance from the centroid to the x axis is

$$Y_\text{c} = 30\,\text{mm} + 30\,\text{mm} = 60\,\text{mm}$$

Therefore, the moment of the area with respect to the x axis is

$$M_x = 2{,}700\,\text{mm}^2 \times 60\,\text{mm} = 162{,}000\,\text{mm}^3$$

We can repeat the same procedure to find the moment of the area of the above triangle with respect to the y axis.

The distance from the centroid to the y axis is

$$X_\text{c} = 2/3(60) + 20\,\text{mm} = 60\,\text{mm},$$

and the moment of area with respect to y axis is

$$M_y = 2{,}700\,\text{mm}^2 \times 60\,\text{mm} = 162{,}000\,\text{mm}^3$$

Example 4.7 Find the moment of the rectangular area shown (Fig. 4.13) with respect to the x and y axes.

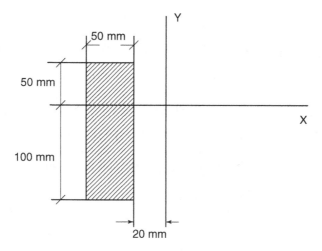

Fig. 4.13

Solution The area of the rectangle is

$$A = 50\,\text{mm} \times 150\,\text{mm} = 7,500\,\text{mm}^2$$

The moment of the area with respect to the x axis is

$$M_x = \left(7,500\,\text{mm}^2\right)(-25\,\text{mm}) = -187,500\,\text{mm}^3$$

And the moment of the area with respect to the y axis is

$$M_y = \left(7,500\,\text{mm}^2\right)(-45\,\text{mm}) = -337,500\,\text{mm}^3$$

Example 4.8 Find the moment of the areas of the circle and the quadrant shown in Fig. 4.14 with respect to the x and y axes.

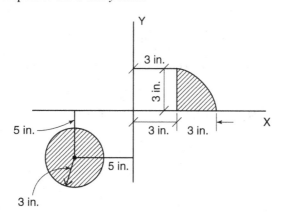

Fig. 4.14

Solution

1. Quadrant: The distance from the centroid to the vertical leg of the quadrant is

$$4r/3\pi = 4(3\,\text{in.})/3\pi = 1.27\,\text{in.}$$

and the distance from the vertical leg to the y axis is

$$X_c = 3\,\text{in.} + 1.27\,\text{in.} = 4.27\,\text{in.}$$

The distance from the centroid to the horizontal is the same as

$$4r/3\pi = 4(3\,\text{in.})/3\pi = 1.27\,\text{in.}$$

Then

$$Y_c = 1.27\,\text{in.}$$

The area of the quadrant is

$$A = \pi r^2/4 = \pi(3\,\text{in.})^2/4 = 7.07\,\text{in.}^2$$

And the moment of the area with respect to the x axis is

$$M_x = (7.07\,\text{in.}^2)(1.27\,\text{in.}) = 8.98\,\text{in.}^3$$

And the moment of the area with respect to the y axis is

$$M_y = (7.07\,\text{in.}^2)(4.27\,\text{in.}^2) = 30.2\,\text{in.}^3$$

2. Circle: The distance from the centroid (center) of the circle to both the x and y axes is 5 in.
   The area of the circle is

$$A = \pi r^2 = \pi(3\,\text{in.})^2 = 28.3\,\text{in.}^2$$

Then the moment of the area of the circle with respect to the x and y axes is the same.

$$M_x = M_y = (28.\,3\,\text{in.}^2)(5\,\text{in.}) = 141.4\,\text{in.}^3$$

4.5   Moment of a Composite Area

Generally, the moment of an area with respect to any axis is equal to the sum of the moments of the divided parts of the entire area with respect to the axis. In other words, if we are looking at the moment of a complicated area, it is logical to divide it into simple areas, find the moment of each of these areas, and then add all the moments to obtain the moment of the entire area. The following examples will clarify this idea.

Example 4.9 Find the moment of the area for the figure shown (Fig. 4.15) with respect to the x and y axes.

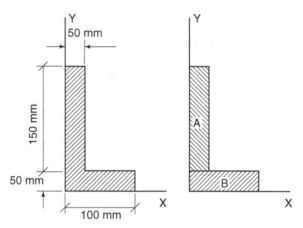

Fig. 4.15

Solution The total area can be divided into two simple areas A and B. Then we find each area with its centroid. Rectangle A has the centroid as follows:

$$X_a = 25\,\text{mm}$$

$$Y_a = 125\,\text{mm}$$

and the area of rectangle A is

$$A_a = 50\,\text{mm} \times 150\,\text{mm} = 7,500\,\text{mm}^2$$

Rectangle B has the centroid as follows:

$$X_b = 50\,\text{mm}$$

$$Y_b = 25\,\text{mm}$$

and the area of rectangle B is

$$A_b = 50\,\text{mm} \times 100\,\text{mm} = 5,000\,\text{mm}^2$$

The moments of the area for rectangle A with respect to the x and y axes are

$$M_x = A_a Y_a = 7,500\,\text{mm}^2 \times 125\,\text{mm} = 937,500\,\text{mm}^3$$

$$M_y = A_a X_a = 7,500\,\text{mm}^2 \times 25\,\text{mm} = 187,500\,\text{mm}^3$$

The moments of the area for rectangle B with respect to the x and y axes are

$$M_x = A_b Y_b = 5,000\,\text{mm}^2 \times 25\,\text{mm} = 12,500\,\text{mm}^3$$

$$M_y = A_b X_b = 5,000\,\text{mm}^2 \times 50\,\text{mm} = 250,000\,\text{mm}^3$$

and the moments of the composite area with respect to the x and y axes are

$$M_x = A_a Y_a + A_b Y_b = 937,500 + 125,000 = 1,062,500\,\text{mm}^3$$

$$M_y = A_a X_a + A_b X_b = 187,500 + 250,000 = 437,500\,\text{mm}^3$$

Example 4.10 Find the moment of the trapezoidal shape shown in Fig. 4.16 with respect to the x and y axes.

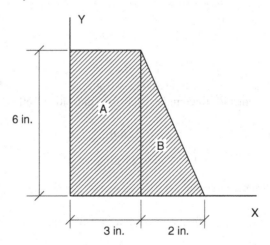

Fig. 4.16

The total area can be divided into two simple areas A and B. Then we find each area and its centroid. Rectangle A has the centroid as follows:

$$X_a = 1.5 \text{ in.}$$

$$Y_a = 3.0 \text{ in.}$$

and the area of rectangle A is

$$A_a = 3 \text{ in.} \times 6 \text{ in.} = 18 \text{ in.}^2$$

Triangle B has the centroid as follows:

$$X_b = 1/3(2 \text{ in.}) + 3 \text{ in.} = 3.67 \text{ in.}$$

$$Y_b = 1/3(6 \text{ in.}) = 2 \text{ in.}$$

and the area of triangle B is

$$A_a = (2 \text{ in.} \times 6 \text{ in.})/2 = 6 \text{ in.}^2$$

The moments of the area for rectangle A with respect to the x and y axes are

$$M_x = A_a Y_a = (18 \text{ in.}^2)(3.0 \text{ in.}) = 54 \text{ in.}^3$$

$$M_y = A_a X_a = (18 \text{ in.}^2)(1.5 \text{ in.}) = 27 \text{ in.}^3$$

The moments of the area for rectangle B with respect to the x and y axes are

$$M_x = A_b Y_b = (6 \text{ in.}^2)(2 \text{ in.}) = 12 \text{ in.}^3$$

$$M_y = A_b X_b = (6 \text{ in.}^2)(3.67 \text{ in.}) = 22 \text{ in.}^3$$

and the moments of the composite area with respect to the x and y axes are

$$M_x = A_a Y_a + A_b Y_b = 54 \text{ in.}^3 + 12 \text{ in.}^3 = 66 \text{ in.}^3$$

$$M_y = A_a X_a + A_b X_b = 27 \text{ in.}^3 + 22 \text{ in.}^3 = 49 \text{ in.}^3$$

Example 4.11 Find the moment of an area for the beam cross section shown (Fig. 4.17) with respect to the x and y axes.

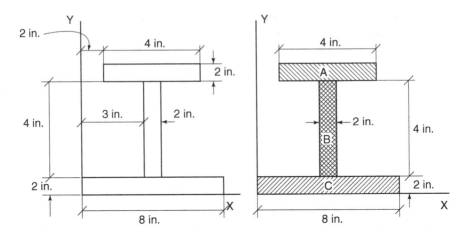

Fig. 4.17

Solution The total area can be divided into three simple areas A, B, and C. Then we find each area with its centroid.

Rectangle A has the centroid as follows:

$$X_a = 4\,\text{in.}$$

$$Y_a = 7\,\text{in.}$$

and the area of rectangle A is

$$A_a = 2\,\text{in.} \times 4\,\text{in.} = 8\,\text{in.}^2$$

Rectangle B has the centroid as follows:

$$X_b = 4\,\text{in.}$$

$$Y_b = 4\,\text{in.}$$

and the area of rectangle B is

$$A_b = (2\,\text{in.} \times 4\,\text{in.}) = 8\,\text{in.}^2$$

Rectangle C has the centroid as follows

$$X_c = 4\,\text{in.}$$

$$Y_c = 1\,\text{in.}$$

And the area of rectangle C is

$$A_c = (2\,\text{in.} \times 8\,\text{in.}) = 16\,\text{in.}^2$$

Notice that areas A, B, and C are all symmetrical with respect to the vertical line passing through the centroid of the entire area.

The moments of the area for rectangle A with respect to the x and y axes are

$$M_x = A_a Y_a = (8\,\text{in.}^2)(7\,\text{in.}) = 56\,\text{in.}^3$$

$$M_y = A_a X_a = (8\,\text{in.}^2)(4\,\text{in.}) = 32\,\text{in.}^3$$

The moments of the area for rectangle B with respect to the x and y axes are

$$M_x = A_b Y_b = (8\,\text{in.}^2)(4\,\text{in.}) = 32\,\text{in.}^3$$

$$M_y = A_b X_b = (8\,\text{in.}^2)(4\,\text{in.}) = 32\,\text{in.}^3$$

The moments of the area for rectangle C with respect to the x and y axes are

$$M_x = A_c Y_c = (16\,\text{in.}^2)(1\,\text{in.}) = 16\,\text{in.}^3$$

$$M_y = A_c X_c = (16\,\text{in.}^2)(4\,\text{in.}) = 64\,\text{in.}^3$$

and the moments of the composite area with respect to the x and y axes are

$$M_x = A_a Y_a + A_b Y_b + A_c Y_c = 56\,\text{in.}^3 + 32\,\text{in.}^3 + 16\,\text{in.}^3 = 104\,\text{in.}^3$$

$$M_y = A_a X_a + A_b X_b + A_c X_c = 32\,\text{in.}^3 + 32\,\text{in.}^3 + 64\,\text{in.}^3 = 128\,\text{in.}^3$$

Example 4.12 Find the moment of the composite area shown (Fig. 4.18) with respect to x and y axes.

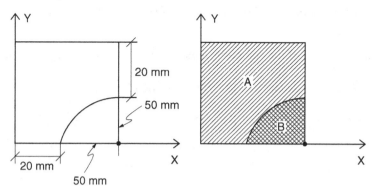

Fig. 4.18

Solution The total area can be divided into two simple areas A and B. Then we find each area with its centroid. Rectangle A has the centroid as follows:

$$X_a = 35\,\text{mm}$$

$$Y_a = 35\,\text{mm}$$

and the area of rectangle A is

$$A_a = 70\,\text{mm} \times 70\,\text{mm} = 4,900\,\text{mm}^2$$

Quadrant B has the centroid as follows:

$$X_b = 70\,\text{mm} - 4r/3\pi = 70 \ - 4 \times 50\,\text{mm}/3\pi = 48.78\,\text{mm}$$

$$Y_b = 4r/3\pi = 21.2\,\text{mm}$$

and the area of quadrant B is negative:

$$A_b = -\pi r^2/4 \ = -\pi(50\,\text{mm})^2/4 = -1,963.5\,\text{mm}^2$$

The moments of the area for rectangle A with respect to the x and y axes are

$$M_x = A_a Y_a = 4,900\,\text{mm}^2 \times 35\,\text{mm} \ = 171,500\,\text{mm}^3$$

$$M_y = A_a X_a = 4,900\,\text{mm}^2 \times 35\,\text{mm} \ = 171,500\,\text{mm}^3$$

The moments of the area for quadrant B with respect to the x and y axes are

$$M_x = A_b Y_b = -1,963.5\,\text{mm}^2 \times 21.2\,\text{mm} = -41,626.2\,\text{mm}^3$$

$$M_y = A_b X_b = -1,963.5\,\text{mm}^2 \times 48.78\,\text{mm} \ = -95,779.5\,\text{mm}^3$$

and the moments of the composite area with respect to the x and y axes are

$$M_x = A_a Y_a + A_b Y_b = 171,500 \ - 41,626.2 = 129,873.8\,\text{mm}^3$$

$$M_y = A_a X_a + A_b X_b = 171,500 - 95,779.5 \ = 75,720.5\,\text{mm}^3$$

Practice Problems

1. Find the moment of the area shown (Fig. 4.19) with respect to the x axis.

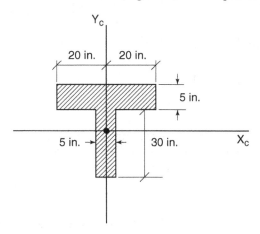

Fig. 4.19

2. Find the moment of the area shown (Fig. 4.20) with respect to the x and y axes.

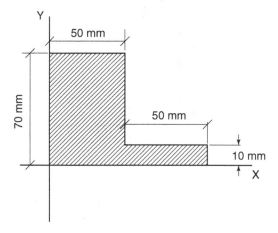

Fig. 4.20

3. Find the moment of the area shown (Fig. 4.21) with respect to the x and y axes.

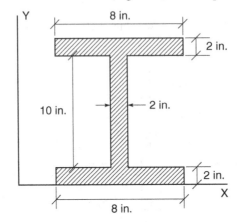

Fig. 4.21

4. Find the moment of the area shown (Fig. 4.22) with respect to the x and y axes.

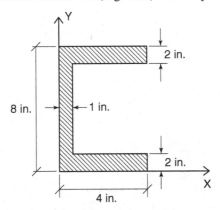

Fig. 4.22

5. Find the moment of the area shown (Fig. 4.23) with respect to the x and y axes.

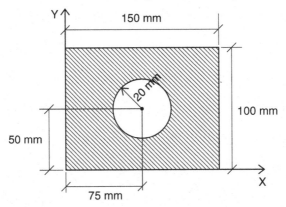

Fig. 4.23

4.6   Centroids of Composite Areas

To find the centroid of a composite area, we must calculate the moment of the composite area as shown in earlier examples. That means we have to follow the approach by dividing the composite area into simple areas, and then find the moment of each simple area. Once we add the moments of these areas together, we get the moment of the composite area with respect to the x or y axis.

Now we can locate the centroid of a composite area, given the moment of a composite area and also the total area. The rest of the calculation is simply a basic algebraic operation. The process is shown in the following examples.

Example 4.13 Find the centroid of the T-section shown in Fig. 4.24.

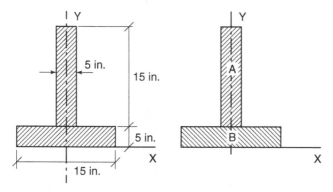

Fig. 4.24

Solution By symmetry, $X_c = 0$, since the y axis is the axis of symmetry. Then $M_y = 0$. The centroid of rectangle A is

$$Y_a = 7.5\,\text{in.} + 5\,\text{in.} = 12.5\,\text{in.}$$

and the area of rectangle A is

$$A_a = 5\,\text{in.} \times 15\,\text{in.} = 75\,\text{in.}^2$$

The centroid of rectangle B is

$$Y_b = (1/2)(5\,\text{in.}) = 2.5\,\text{in.}$$

and the area of rectangle B is

$$A_b = 5\,\text{in.} \times 15\,\text{in.} = 75\,\text{in.}^2$$

Now we can calculate moment of the composite area:

$$M_x = A_a Y_a + A_b Y_b = (75\,\text{in.}^2)(12.5\,\text{in.}) + (75\,\text{in.}^2)(2.5\,\text{in.}) = 1,125\,\text{in.}^3$$

and the total area is

$$A_{\text{total}} = A_a + A_b = 75\,\text{in.}^2 + 75\,\text{in.}^2 = 150\,\text{in.}^2$$

The centroid of the composite area is

$$Y_c = M_x/A_{\text{total}} = 1,125\,\text{in.}^3/\,150\,\text{in.}^2 = 7.5\,\text{in.}$$

Example 4.14 Find the centroid of the trapezoid area shown (Fig. 4.25) with respect to the x and y axes.

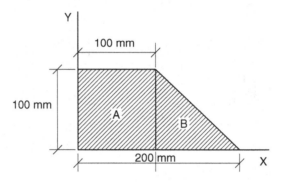

Fig. 4.25

Solution The total area can be divided into two simple areas A and B. Then we find each area with its centroid.

Rectangle A has the centroid as follows:

$$X_a = 50\,\text{mm}$$

$$Y_a = 50\,\text{mm}$$

and the area of rectangle A is

$$A_a = 100\,\text{mm} \times 100\,\text{mm} = 10,000\,\text{mm}^2$$

Triangle B has the centroid as follows:

$$X_b = 1/3(100\,\text{mm}) + 100\,\text{mm} = 133.33\,\text{mm}$$

$$Y_b = 1/3(100\,\text{mm}) = 33.3\,\text{mm}$$

and the area of triangle B is

$$A_b = (100\,\text{mm} \times 100\,\text{mm})/2 = 5,000\,\text{mm}^2$$

The moments of the area for rectangle A with respect to the x and y axes are

$$M_x = A_a Y_a = (10,000\,\text{mm}^2)(50\,\text{mm}) = 500,000\,\text{mm}^3$$

$$M_y = A_a X_a = (10,000\,\text{mm}^2)(50\,\text{mm}) = 500,000\,\text{mm}^3$$

The moments of the area for rectangle B with respect to the x and y axes are

$$M_x = A_b Y_b = (5,000\,\text{mm}^2)(33.3\,\text{mm}) = 166,500\,\text{mm}^3$$

$$M_y = A_b X_b = (5,000\,\text{mm}^2)(133.33\,\text{mm}) = 666,650\,\text{mm}^3$$

and the total moments of the composite area with respect to the x and y axes are

$$M_x = A_a Y_a + A_b Y_b = 500,000\,\text{mm}^3 + 166,500\,\text{mm}^3 = 666,500\,\text{mm}^3$$

$$M_y = A_a X_a + A_b X_b = 500,000\,\text{mm}^3 + 666,650\,\text{mm}^3 = 1,166,650\,\text{mm}^3$$

And the total area is

$$A_{\text{total}} = A_a + A_b = 10,000\,\text{mm}^2 + 5,000\,\text{mm}^2 = 15,000\,\text{mm}^2$$

and the coordinates of the centroid of a trapezoid area are

$$X_c = M_y/A_{\text{total}} = 1,166,650\,\text{mm}^3/15,000\,\text{mm}^2 = 77.8\,\text{mm}$$

$$Y_c = M_x/A_{\text{total}} = 666,500\,\text{mm}^3/15,000\,\text{mm}^2 = 44.4\,\text{mm}$$

Example 4.15 Find the centroid of angle section (L) shown (Fig. 4.26) with respect to the x and y axes.

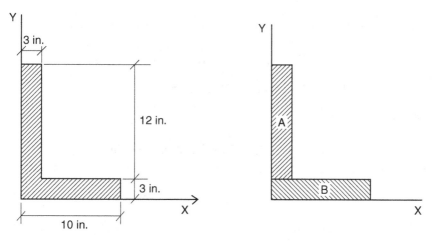

Fig. 4.26

Solution The total area can be divided into two simple areas A and B. Then we find each area with its centroid.

Rectangle A has the centroid as follows:

$$X_a = 1.5\,\text{in.}$$

$$Y_a = 9\,\text{in.}$$

and the area of rectangle A is

$$A_a = 3\ \text{in.} \times 12\,\text{in.} = 36\,\text{in.}^2$$

Rectangle B has the centroid as follows:

$$X_b = 5\,\text{in.}$$

$$Y_b = 1.5\,\text{in.}$$

and the area of the rectangle B is

$$A_b = 3\,\text{in.} \times 10\,\text{in.} = 30\,\text{in.}^2$$

The moments of the area for rectangle A with respect to the x and y axes are

$$M_x = A_a Y_a = 36\,\text{in.}^2 \times 9\,\text{in.} = 324\,\text{in.}^3$$

$$M_y = A_a X_a = 36\,\text{in.}^2 \times 1.5\,\text{in.} = 54\,\text{in.}^3$$

The moments of the area for rectangle B with respect to x and y axes are

$$M_x = A_b Y_b = 30\,\text{in.}^2 \times 1.5\,\text{in.} = 45\,\text{in.}^3$$

$$M_y = A_b X_b = 30\,\text{in.}^2 \times 5\,\text{in.} = 150\,\text{in.}^3$$

and the total moments of the composite area with respect to the x and y axes are

$$M_x = A_a Y_a + A_b Y_b = 324\,\text{in.}^3 + 45\,\text{in.}^3 = 369\,\text{in.}^3$$

$$M_y = A_a X_a + A_b X_b = 54\,\text{in.}^3 + 150\,\text{in.}^3 = 204\,\text{in.}^3$$

And the total area is

$$A_{\text{total}} = A_a + A_b = 36\,\text{in.}^2 + 30\,\text{in.}^2 = 66\,\text{in.}^2$$

The two coordinates of the centroid of the L section area are

$$X_c = M_y/A_{total} = 204\,in.^3\,/\,66\,in.^2 = 3.1\,in.$$

$$Y_c = M_x\,/A_{total} = 369\,in.^3/66\,in.^2 = 5.6\,in.$$

Practice Problems

1. Find the centroid W of the beam section shown (Fig. 4.27) with respect to the x and y axes.

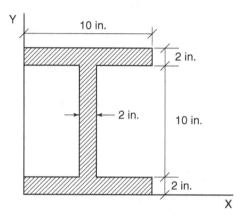

Fig. 4.27

2. Find the centroid of the angle section (L) shown (Fig. 4.28) with respect to the x and y axes.

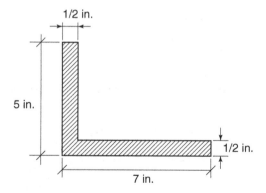

Fig. 4.28

3. Find the centroid of the channel section shown (Fig. 4.29) with respect to the x and y axes.

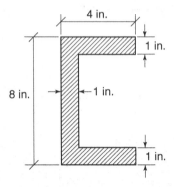

Fig. 4.29

4. Find the centroid of the composite area shown (Fig. 4.30) with respect to the x and y axes.

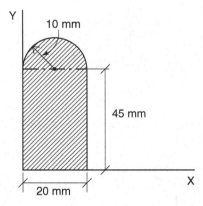

Fig. 4.30

5. Find the centroid of the composite area shown (Fig. 4.31) with respect to the x and y axes.

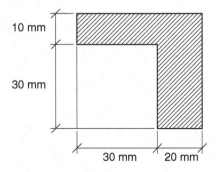

Fig. 4.31

Chapter Summary

1. The center of gravity of a body is an imaginary point at which all its weight passes through.
2. There is no center of gravity of an area, for an area does not have weight.
3. The centroid of an area is used for the two dimensional shape.
4. The location of the centroid is very important in structural design when the body undergoes a bending, and bending neutral axis becomes centroidal axis.
5. Moment of an area is the product of the amount of the area and the distance of the centroid of the area from the axis.
6. The moment of a composite area is equal to sum of the moments of divided area with respect to the axis.
7. To find the coordinates (x, y) of the centroid of the composite area, we simply divide the moment of the composite area respect to the axis by the total area.

Review Questions

1. What is the center of gravity of a body?
2. What is the centroid of an area?
3. What is the moment of an area with respect of an axis?
4. What is a composite area?
5. What is the moment of a composite area, and how to calculate it with respect to the x and y axes?
6. How to calculate the centroid of a composite area with respect to the x and y axes?

Problems

1. Find the centroid of the triangle shown (Fig. 4.32) with respect to the x and y axes.

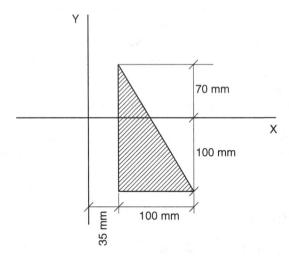

Fig. 4.32

2. Find the centroid of the rectangle shown (Fig. 4.33) with respect to the x and y axes.

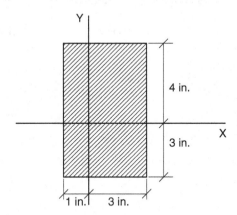

Fig. 4.33

3. Find the centroid of the quadrant shown (Fig. 4.34) with respect to the x and y axes.

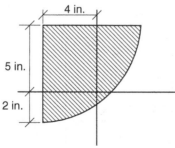

4 in.

5 in.

2 in.

Fig. 4.34

4. Find the centroid of the beam section shown (Fig. 4.35) with respect to the x and y axes.

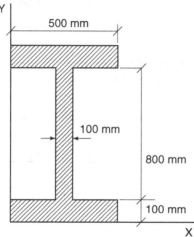

Y

500 mm

100 mm

800 mm

100 mm

X

Fig. 4.35

5. Find the centroid of the composite area shown (Fig. 4.36) with respect to the x and y axes.

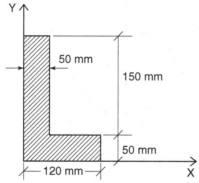

Y

50 mm

150 mm

50 mm

120 mm

X

Fig. 4.36

6. Find the centroid of the composite area shown (Fig. 4.37) with respect to the
   x and y axes.

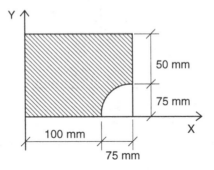

7. Find the centroid of the composite area shown (Fig. 4.38) with respect to the
   x and y axes.

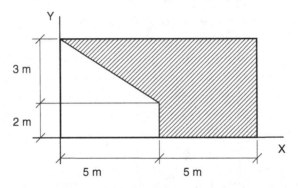

8. Find the centroid of the composite area shown (Fig. 4.39) with respect to the
   x and y axes.

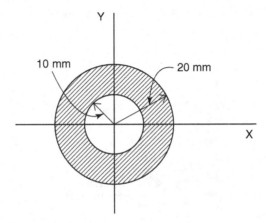

9. Find the centroid of the hollow shape shown (Fig. 4.40) with respect to the x and y axes.

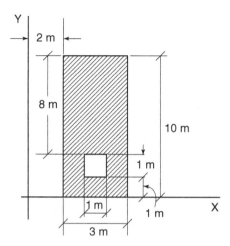

Fig. 4.40

Moment of Inertia

5

Overview
When we study equations in mechanics of materials, we use some of the properties of sections. One of these properties that depends on the size and shape of the member is called the moment of inertia. Moment of inertia is a mathematical property of an area that controls resistance to bending, buckling, or rotation of the member.

Learning Objectives
Upon completion of this chapter, you will be able to calculate the moment of inertia of an area. Also, you will learn about of one the important properties of an area. This property is used extensively in structural engineering designs dealing with bending or rotation of the members. Your knowledge, application, and problem solving skills will be determined by your performance on the chapter test.

Upon completion of this chapter, you will be able to:

- *Define the moment of an area*
- *Define and calculate the moment of inertia of a simple area*
- *Define and calculate the moment of inertia using the parallel axis theorem with respect to an axis other than the centroidal axis*
- *Calculate the moment of inertia of a composite area*

5.1   Moment of Inertia

The moment of inertia of an area is the capacity of a cross section to resist bending or buckling. It represents a mathematical concept that is dependent on the size and shape of the section of the member. The bending axis of a member is also the centroidal axis; therefore, the ability to locate the centroid of a shape is closely associated with moment of inertia. Engineers use the moment of inertia to determine the state of stress in a section, and determine the amount of deflection in a beam.

© Springer International Publishing Switzerland 2015
P. Ghavami, *Mechanics of Materials*, DOI 10.1007/978-3-319-07572-3_5

The definition of the moment of inertia of an area can be thought of as the sum of the products of all the small areas and the squares of their distances from the axis being considered. This gives

$$\sum ay^2 \quad \text{with respect to the } x \text{ axis}$$

or

$$\sum ax^2 \quad \text{with respect to the } y \text{ axis}$$

If we represent the moment of inertia by the letter I, then the moment of inertia with respect to the x axis is

$$I_x = \sum ay^2$$

Similarly, the moment of inertia with respect to the y axis is

$$I_y = \sum ax^2$$

Units

Moment of inertia is expressed in units of length to the fourth power. Although dimensionally speaking it seems unusual, it is just a mathematical abstract and is an important property in the design of beams and columns. We will see in the following examples the methods of calculating the moment of inertia for a given beam section subjected to bending. If we choose the unit of length as in., then the unit of the moment of inertia is

$$\text{in.}^2 \times \text{in.}^2 = \text{in.}^4$$

In the metric system, if the unit of length taken is mm, then the unit of the moment of inertia is

$$\text{mm}^2 \times \text{mm}^2 = \text{mm}^4$$

5.2   Moment of Inertia of Simple Areas

Using calculus and integrating equations for an area, we will be able to extract the exact values for the moment of inertia. Derivation of the moment of inertia formulas for most commonly used shapes such as rectangle, triangle, and circle are given in Table 5.1. The *radius of gyration* (r) of an area in the table represents the distance from the moment of inertia axis at which the entire area could be considered without changing its moment of inertia. The radius of gyration is a property of a shape's area and is mostly used in column design. It is expressed as

Table 5.1 Properties of areas

Shape	Area (A)	Moment of inertia (I)	Radius of gyration (r)	Polar moment of inertia (J)
Rectangle	$A = bh$	$I_{xo} = \dfrac{bh^3}{12}$ $I_{yo} = \dfrac{hb^3}{12}$ $I_x = \dfrac{bh^3}{3}$	$r_{xo} = \dfrac{h}{\sqrt{12}}$ $r_{yo} = \dfrac{b}{\sqrt{12}}$ $r_x = \dfrac{h}{\sqrt{3}}$	$J_{CG} = \dfrac{bh}{12}\left(h^2 + b^2\right)$
Triangle	$A = \dfrac{bh}{2}$	$I_{xo} = \dfrac{bh^3}{36}$ $I_x = \dfrac{bh^3}{12}$	$r_{xo} = \dfrac{h}{\sqrt{18}}$ $r_x = \dfrac{h}{\sqrt{6}}$	

(continued)

Table 5.1 (continued)

Shape	Area (A)	Moment of inertia (I)	Radius of gyration (r)	Polar moment of inertia (J)
Circle	$A = \dfrac{\pi d^2}{4}$ $= 0.7854d^2$	$I_{xo} = I_{yo} = \dfrac{\pi d^4}{64}$	$r_{xo} = r_{yo} = \dfrac{d}{4}$	$J_{CG} = \dfrac{\pi d^4}{32}$
Semicircle	$A = \dfrac{\pi R^2}{2}$ $= 1.571R^2$	$I_{xo} = 0.1098R^4$ $I_{yo} = I_x = \dfrac{\pi R^4}{8}$ $= 0.3927R^4$	$r_{xo} = 0.264R$ $r_{yo} = r_x = \dfrac{R}{2}$	$J_{CG} = I_{xo} + I_{yo}$ $= 0.5025R^4$ $J_o = \dfrac{\pi R^4}{4}$
Hollow Circle	$A = \dfrac{\pi(d^2 - d_1^2)}{4}$ $= 0.7854(d^2 - d_1^2)$	$I_{xo} = \dfrac{\pi(d^4 - d_1^4)}{64}$ $I_{yo} = I_{xo}$	$r_{xo} = \dfrac{\sqrt{d^2 + d_1^2}}{4}$ $r_{yo} = r_{xo}$	$J_{CG} = \dfrac{\pi(d^4 - d_1^4)}{32}$

	A	I	r	J
Hollow Rectangle	$A = bd - b_1 d_1$	$I_{xo} = \dfrac{bd^3 - b_1 d_1^3}{12}$ $I_{yo} = \dfrac{db^3 - d_1 b_1^3}{12}$	$r_{xo} = \sqrt{\dfrac{bd^3 - b_1 d_1^3}{12A}}$ $r_{yo} = \sqrt{\dfrac{db^3 - d_1 b_1^3}{12A}}$	$J_{CG} = I_{xo} + I_{yo}$
Quarter-Circle	$A = \dfrac{\pi R^2}{4}$	$I_{xo} = I_{yo} = 0.0549R^4$ $I_x = \dfrac{\pi R^4}{16}$	$r_{xo} = r_{yo} = 0.2644R$ $r_x = 0.5R$	$J_{CG} = 0.1098R^4$

$$r = \sqrt{I/A}$$

J is called the *polar moment of inertia* and is defined as the moment of inertia with respect to an axis perpendicular to the plane of the area. In this case, it could be the Z–Z axis perpendicular to the X–Y plane.

Therefore,

$$J = \sum ar^2$$

where
$r^2 = x^2 + y^2$ for any right triangle, and r is the moment arm.

It must be noted here that these formulas are only applicable for an area's centroidal axis. However, to find the moment of inertia for an axis other than the centroidal axis, a different approach which will be discussed later in this chapter is used.

Example 5.1 Find the value of the moment of inertia for the 6×14-in. rectangular beam shown (Fig. 5.1) with respect to the axis passing through its centroid and parallel to the base.

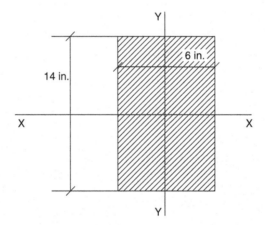

Fig. 5.1

Solution The formula used for moment of inertia of a rectangle with respect to the axis passing through the centroid and parallel to the base is given in Table 5.1. Then

$$I_{xc} = bh^3/12$$

$b = 6$ in. and is the width of the rectangle
$h = 14$ in. and is the depth of the rectangle.
Then the calculation of the moment of inertia with respect to the axis parallel to base is

$$I_{xc} = (6\,\text{in.})(14\,\text{in.})^3/12 = 1,372\,\text{in.}^4$$

Example 5.2 Find the moment of inertia of the triangular cross section shown (15-in. base and 9-in. height) with respect to an axis passing through its centroid and parallel to the base (Fig. 5.2).

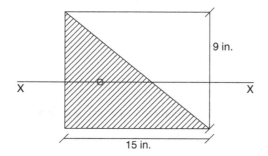

Fig. 5.2

Solution The moment of inertia of a triangle with respect to the axis passing through the centroid and parallel to the base is given in Table 5.1.

$$I_{xc} = bh^3/36$$

$b = 15$ in. and is the base of the triangle
$h = 9$ in. and is the height of the triangle
Then the calculation of the moment of inertia with respect to the axis passing through the centroid and parallel to the base is

$$I_{xc} = (15\,\text{in.})(9\,\text{in.})^3/36 = 303.75\,\text{in.}^4$$

Example 5.3

Find the moment of inertia of the circular cross section shown with a diameter of 25 mm with respect to an axis passing through its centroid (Fig. 5.3).

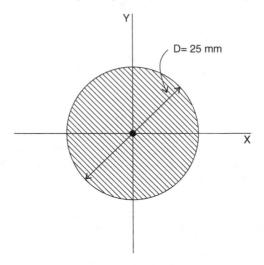

Fig. 5.3

Solution The formula used for moment of inertia of a circular section is given in Table 5.1. Then

$$I_{xc} = \pi d^4 / 64$$

where d (diameter) of the circle $= 25$ mm
Using the above formula, we get

$$I_{xc} = \pi (25 \, \text{mm})^4 / 64 = 19,174.76 \, \text{mm}^4$$

Example 5.4

Find the moment of inertia of the rectangular section shown with respect to the x and y axes passing through the centroid (Fig. 5.4).

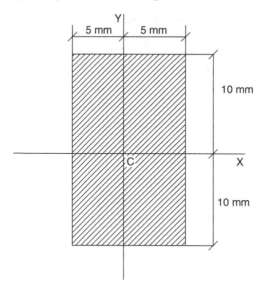

Fig. 5.4

Solution   We calculate I_x and I_y in the same manner we did earlier.

The formula used for moment of inertia I_x of a rectangle is given in Table 5.1. Then

$$I_{xc} = b h^3 / 12$$

$b = 10$ mm and is the width of the rectangle
$h = 20$ mm and is the depth of the rectangle
The calculation of the moment of inertia with respect to the x axis is

$$I_{xc} = (10\,\text{mm})(20\,\text{mm})^3 / 12 = 6,666.67\,\text{mm}^4$$

And for I_y, we have

$$I_{yc} = h b^3 / 12$$

Substituting $h = 20$ mm and $b = 10$ mm into the above formula, we get

$$I_{yc} = (20\,\text{mm})(10\,\text{mm})^3 / 12 = 1,666.67\,\text{mm}^4$$

Practice Problems

In the following problems, find the moment of inertia with respect to the centroidal axis of the beam sections shown.

1. The rectangular cross section shown in Fig. 5.5.

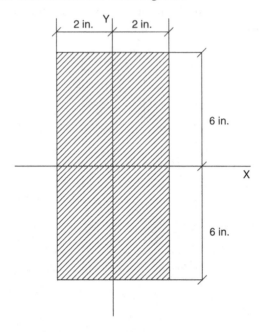

Fig. 5.5

2. The triangular cross section shown in Fig. 5.6.

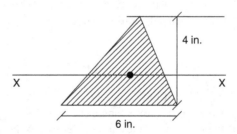

Fig. 5.6

3. The circular cross section shown in Fig. 5.7.

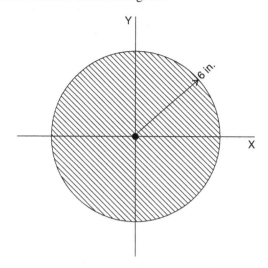

Fig. 5.7

4. The semicircular cross section shown in Fig. 5.8.

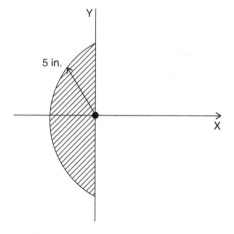

Fig. 5.8

5. The quadrant cross section shown in Fig. 5.9.

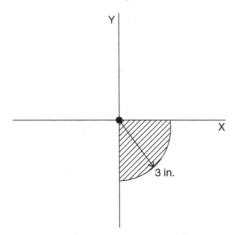

Fig. 5.9

6. The rectangular cross section shown in Fig. 5.10.

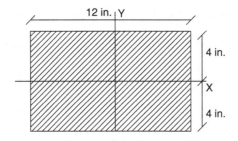

Fig. 5.10

7. The right triangular shape shown in Fig. 5.11.

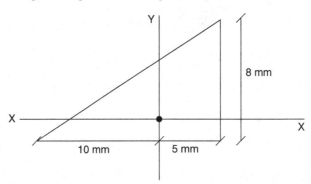

Fig. 5.11

8. The semicircular cross section shown in Fig. 5.12.

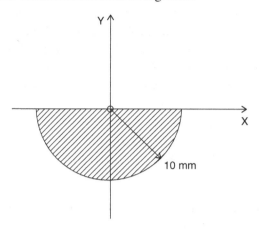

Fig. 5.12

5.3   **Parallel Axis Theorem**

When it is necessary to find the moment of inertia of an area with respect to any axis other than the centroidal axis, the parallel axis theorem must be used. This theorem simply says that the moment of inertia with respect to any axis parallel to its centroidal axis is equal to its centroidal moment of inertia plus the area times the square of the distance between two axes. This theorem is also called the transfer formula (Fig. 5.13).

In equation form:

$$I_x = I_{xc} + Ad^2$$

or

$$I_y = I_{yc} + Ad^2$$

Here I_x and I_y are the moments of inertia of an area with respect to the x and y axes, respectively. I_{xc} and I_{yc} are the moments of inertia with respect to the centroidal axis. A is the area of the cross section, and d is the perpendicular distance between the two parallel axes.

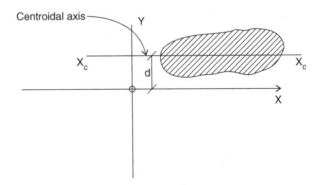

Fig. 5.13

Example 5.5 Find the moment of inertia of the rectangular cross section shown in Fig. 5.14 with respect to the x axis.

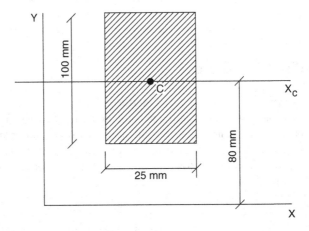

Fig. 5.14

Solution From Table 5.1, the formula for a rectangular cross section with respect to its centroidal axis is

$$I_{xc} = bh^3/12$$

or

$$I_{xc} = (25\,\text{mm})(100\,\text{mm})^3/12 = 2,083,333\,\text{mm}^4$$

The area of the rectangle is

$$A = (25\,\text{mm})(100\,\text{mm}) = 2,500\,\text{mm}^2$$

Therefore,

$$I_x = I_{xc} + Ad^2$$

or

$$I_x = 2,083,333\,\text{mm}^4 + (2,500\,\text{mm}^2)(80\,\text{mm})^2 = 18,083,333\,\text{mm}^4$$

Example 5.6 Find the moment of inertia of the triangular cross section shown in Fig. 5.15 with respect to the x axis.

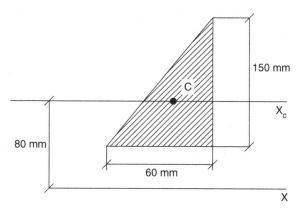

Fig. 5.15

Solution From previous examples, we have

$$I_{xc} = bh^3/36 = (60\,\text{mm})(150\,\text{mm})^3/36 = 5,625,000\,\text{mm}^4$$

$$A = (60\,\text{mm})(150\,\text{mm})/2 = 4,500\,\text{mm}^2$$

$$d = 80\,\text{mm}$$

Therefore,

$$I_x = I_{xc} + Ad^2 = 5,625,000\,\text{mm}^4 + (4,500\,\text{mm}^2)(80\,\text{mm})^2 = 34,425,000\,\text{mm}^4$$

Example 5.7

Find the moment of inertia of the circular cross section shown in Fig. 5.16 with respect to the x axis.

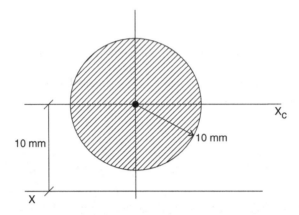

Fig. 5.16

Solution From Table 5.1, the formula for a circular cross section with respect to its centroidal axis is

$$I_{xc} = \pi d^4/64 = \pi (20\,\text{mm})^4/64 = 7,853.98\,\text{mm}^4$$

$$A = \pi r^2 = \pi (10\,\text{mm})^2 = 314.2\,\text{mm}^2$$

$$d = 10\,\text{mm}$$

Then

$$I_x = I_{xc} + Ad^2$$

$$I_x = 7,853.98\,\text{mm}^4 + (314.2\,\text{mm}^2)(10\,\text{mm})^2 = 39,273.98\,\text{mm}^4$$

Example 5.8
Find the moment of inertia of the triangular cross section shown in Fig. 5.17 with respect to the y axis.

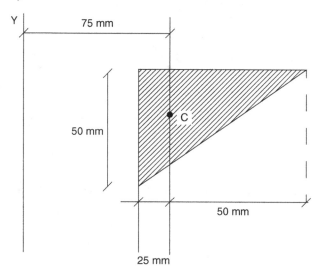

Fig. 5.17

Solution   From Table 5.1, the formula for a triangular cross section with respect to its centroidal axis is

$$I_{xc} = bh^3/36 = (50\,\text{mm})(75\,\text{mm})^3/36 = 585{,}937.5\,\text{mm}^4$$

$$A = bh/2 = (50\,\text{mm})(75\,\text{mm})/2 = 1{,}875\,\text{mm}^2$$

$$d = 75\,\text{mm}$$

Then

$$I_x = I_{xc} + Ad^2$$

$$I_x = 585{,}937.5\,\text{mm}^4 + \left(1{,}875\,\text{mm}^2\right)(75\,\text{mm})^2 = 11{,}132{,}812.5\,\text{mm}^4$$

Practice Problems
Using the parallel axis theorem, find the moment of inertia of the following cross sections with respect to the x or y axis, or both axes parallel to the centroidal axis.

1. Triangular cross section shown in Fig. 5.18 with respect to the x and y axes.

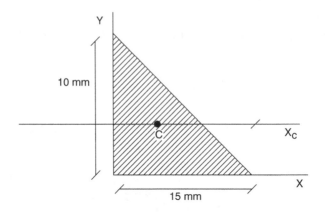

Fig. 5.18

2. Rectangular cross section shown in Fig. 5.19 with respect to the *x* and *y* axes.

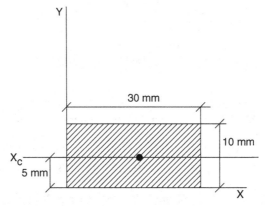

Fig. 5.19

3. Rectangular cross section shown in Fig. 5.20 with respect to the *x* axis.

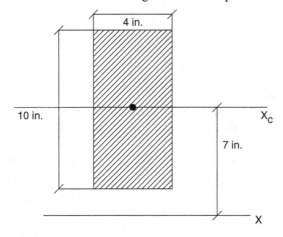

Fig. 5.20

4. Circular cross section shown in Fig. 5.21 with respect to the x and y axes.

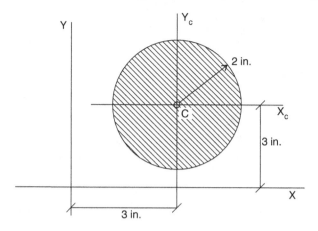

Fig. 5.21

5. Semicircular cross section shown in Fig. 5.22 with respect to the x and y axes.

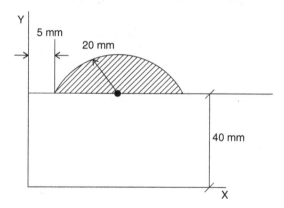

Fig. 5.22

6. Triangular cross section shown in Fig. 5.23 with respect to the y axis.

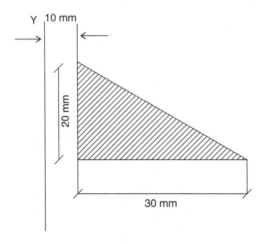

Fig. 5.23

7. Quadrant cross section shown in Fig. 5.24 with respect to x and y axes.

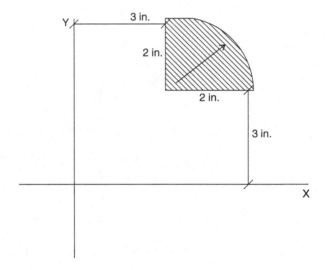

Fig. 5.24

5.4   Moment of Inertia of Composite Areas

As we discussed earlier in Chap. 4, a composite area is an area that can be divided into simple areas. Composite areas are frequently used in the construction industry, such as in I-beams and column cross sections, or other structural members. To calculate the moment of inertia for a composite area, we simply apply the same type of formulations and methodology that we used for simple areas, except that we must add the moment of inertia of each part to come up with the total moment of inertia of the entire area, which is the moment of inertia of the composite area.

Example 5.9 Calculate the moment of inertia of the I-beam shown in Fig. 5.25 with respect to the centroidal x axis.

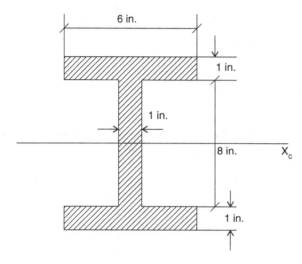

Fig. 5.25

Solution We divide the entire area into two areas, rectangle A with a width of 6 in. and depth of 10 in. and area B consisting of two open spaces at the sides of a vertical web.

For area A:

$$I_{xc} = bh^3/12 = (6\,\text{in.})(10\,\text{in.})^3/12 = 500\,\text{in.}^4$$

For area B:

Two open spaces are equivalent to a rectangular area with a width of 5 in. and depth of 8 in. Then the moment of inertia of the new rectangular area is

$$I_{xc} = bh^3/12 = (5\,\text{in.})(8\,\text{in.})^3/12 = 213.3\,\text{in.}^4$$

So the moment of inertia of the composite area with respect to the x axis is

$$I_x = 500\,\text{in.}^4 - 213.3\,\text{in.}^4 = 286.7\,\text{in.}^4$$

Example 5.10 Calculate the moment of inertia of the hollow circular shape shown in Fig. 5.26 with respect to the centroidal x and y axes.

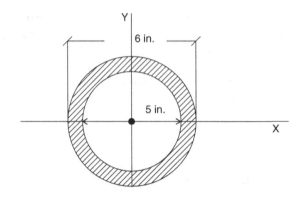

Fig. 5.26

Solution There are two methods for calculating the moment of inertia of a hollow circular area with respect to the centroidal x and y axes.

Method 1 Calculate it directly using the formulation given in Table 5.1.

$$
\begin{aligned}
I_{xc} &= I_{yc} = \pi\left(d_o{}^4 - d_i{}^4\right)/64 \\
&= \pi\left[(6\,\text{in.})^4 - (5\,\text{in.})^4\right]/64 = 32.94\,\text{in.}^4
\end{aligned}
$$

Method 2 The moment of inertia for the outer circle with $d_o = 6$ in. is

$$
I_{xo} = \pi\left(d_o{}^4\right)/64 = \pi(6\,\text{in.})^4/64 = 63.62\,\text{in.}^4
$$

The moment of inertia for the inner circle with $d_i = 5$ in. is

$$
Ix_i = \pi\left(d_i{}^4\right)/64 = \pi(5\,\text{in.})^4\Big]/64 = 30.68\,\text{in.}^4
$$

So the moment of inertia for the composite area with respect to the x axis is

$$
I_x = I_{xo} - Ix_i = 63.62\,\text{in.}^4 - 30.68\,\text{in.}^4 = 32.94\,\text{in.}^4
$$

This matches with the answer from method 1.

Example 5.11 Calculate the moment of inertia for the built-up beam shown in Fig. 5.27 with respect to the centroidal x axis.

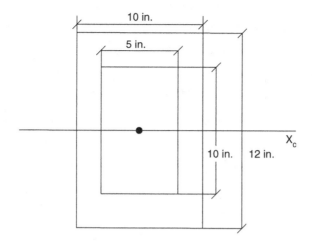

Fig. 5.27

Solution The moment of inertia of the 10×12-in. rectangle is

$$I_{xo} = bh^3/12 = (10\,\text{in.})(12\,\text{in.})^3/12 = 1,440\,\text{in.}^4$$

The moment of inertia of the 5×10-in. rectangle is

$$I_{xi} = bh^3/12 = (5\,\text{in.})(10\,\text{in.})^3/12 = 416.67\,\text{in.}^4$$

So the moment of inertia for the composite area with respect to the x axis is

$$I_x = I_{xo} - I_{xi} = 1,440\,\text{in.}^4 - 416.67\,\text{in.}^4 = 1,023.33\,\text{in.}^4$$

Example 5.12 Calculate the moment of inertia for the channel shown in Fig. 5.28 with respect to the centroidal x axis.

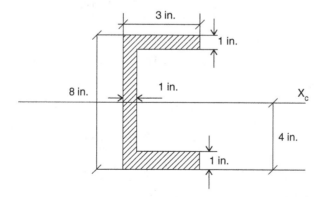

Fig. 5.28

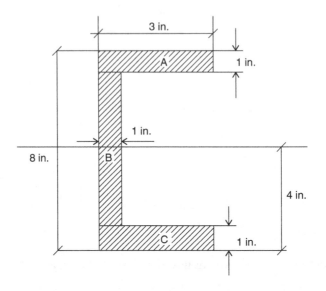

Solution The channel can be divided into three areas A, B, and C. Then we can calculate the moment of inertia of each area with respect to the centroidal x axis of the channel. For this purpose we use the parallel axis theorem discussed earlier. Vertical distance d_y is the distance from the centroidal x axis of each area to the channel centroidal axis (x_c). Then the total moment of inertia of the composite area is

$$I_x = \sum \left(I_{xc} + A d_y^2 \right)$$

For area A:

$$
\begin{aligned}
I_{xc} &= bh^3/12 \\
&= (3\,\text{in.})(1\,\text{in.})^3/12 = 0.25\,\text{in.}^4
\end{aligned}
$$

$$A = 1\,\text{in.} \times 3\,\text{in.} = 3\,\text{in.}^2$$

$$d = 3.5\,\text{in.}$$

For area B:

$$
\begin{aligned}
I_{xc} &= bh^3/12 \\
&= (1\,\text{in.})(6\,\text{in.})^3/12 = 18\,\text{in.}^4
\end{aligned}
$$

$$A = 1\,\text{in.} \times 6\,\text{in.} = 6\,\text{in.}^2$$

$$d = 0$$

For area C:

$$I_{xc} = bh^3/12$$
$$= (3\,\text{in.})(1\,\text{in.})^3/12 = 0.25\,\text{in.}^4$$

$$A = 1\,\text{in.} \times 3\,\text{in.} = 3\,\text{in.}^2$$

$$d = 3.5\,\text{in.}$$

Then

$$\sum I_{xc} = .25(2) + 18 = 18.5\,\text{in.}^4$$
$$\sum \left(Ad_y^2\right) = \left[3(3.5)^2 \times 2\right] = 73.5\,\text{in.}^4$$

Therefore,

$$I_x = \sum I_{xc} + \sum \left(Ad_y^2\right)$$

or

$$I_x = 18.5\,\text{in.}^4 + 73.5\,\text{in.}^4 = 92\,\text{in.}^4$$

Practice Problems

For the following figures, calculate the moment of inertia with respect to the centroidal x axis.

1.

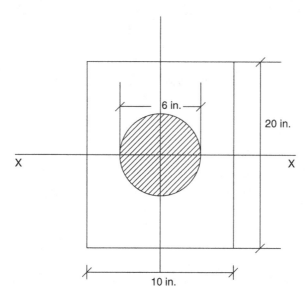

6 in.

20 in.

X   X

10 in.

Fig. 5.29

2.

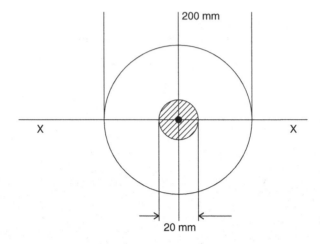

Fig. 5.30

3.

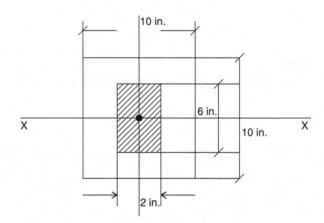

Fig. 5.31

4.

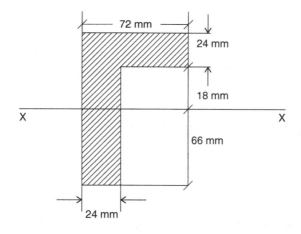

Fig. 5.32

5.

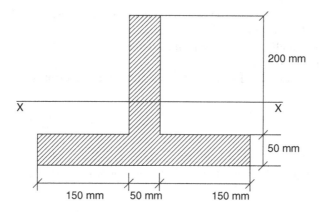

Fig. 5.33

Chapter Summary

The moment of inertia is a property of an area. It is a mathematical quantity that affects the load-carrying capacity of beams and columns. An increase in the moment of inertia with respect to an axis will produce higher resistance to bending forces. The unit of inertia is in.^{4}, or mm^4. Moments of inertia with respect to the centroidal x and y axes are

$$I_{xc} = \sum Ay^2$$

$$I_{yc} = \sum Ax^2$$

The moment of inertia of an area with respect to the axis parallel to the centroidal axis is found using the parallel axis theorem

$$I_{xc} = I_{xc} + Ad^2$$

$$I_y = I_{yc} + Ad^2$$

Review Questions

1. What is the moment of inertia with respect to the centroidal axis parallel to the base?
2. What is the moment of inertia of a circular area with respect to the x and y axes through the centroid?
3. What is the moment of inertia of a triangular shape with respect to the centroidal axis parallel to the base?
4. What is the parallel axis theorem and when can it be used?
5. What is a composite area?
6. State the procedure for how to calculate the moment of inertia of a composite area.

Problems

Find the moment of inertia with respect to the centroidal x axis for the following composite areas.

1. T-section in Fig. 5.34.

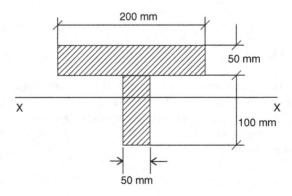

Fig. 5.34

2. Channel section in Fig 5.35.

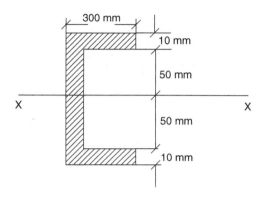

Fig. 5.35

3. Trapezoid area in Fig. 5.36.

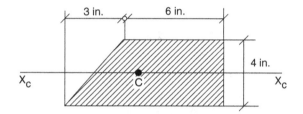

Fig. 5.36

4. Composite area shown in Fig. 5.37.

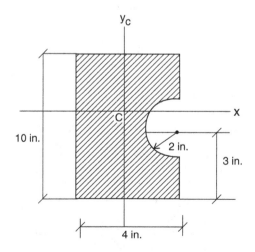

Fig. 5.37

5. Composite area shown in Fig. 5.38.

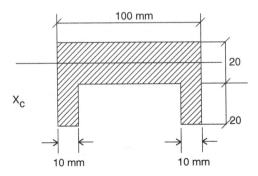

Fig. 5.38

6. Hollow circular composite area shown in Fig. 5.39.

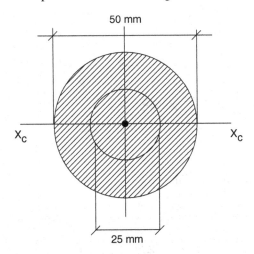

Fig. 5.39

7. Composite shaded area shown in Fig. 5.40.

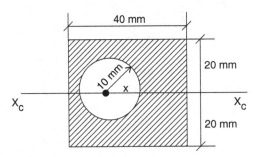

Fig. 5.40

8. Composite shaded area shown in Fig. 5.41.

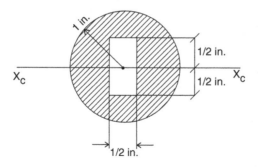

Fig. 5.41

9. Composite area shown in Fig. 5.42.

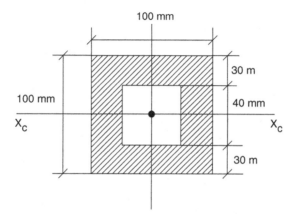

Fig. 5.42

10. Composite area shown in Fig. 5.43.

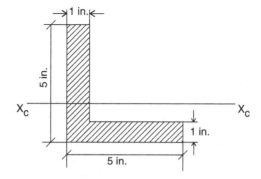

Fig. 5.43

Stress and Strain

6

Overview

Designers of machinery and structures must know the relationship between the external forces acting on elastic bodies and the internal forces developed due to the influence of these forces. In this chapter, we will study new technical terms such as stress, strain, deformation, and modulus of elasticity. These are the concepts that a designer must take into consideration when dealing with elastic bodies. Both the analysis and the design of various elements of a structure are involved the determination of stress and deformation and the selection of proper materials based on the principles of mechanics of materials.

Learning Objectives

Upon completion of this chapter, you will be able to define and compute tensile or compressive stress, shear stress, and allowable stress. Also, you will be able to calculate the deformation (elongation) of elastic bodies subject to external forces and plot the stress–strain diagrams for different materials, and identify the modulus of elasticity based on the stress–strain diagrams of materials. Your knowledge, application, and problem solving skills will be determined by your performance on the chapter test.

Upon completion of this chapter, you will be able to:

- *Define and compute stress under tension or compression*
- *Define shearing stress and allowable stress in building structures*
- *Define and calculate deformation (elongation) of different materials*
- *Define and compute strain under tension or compression*
- *Define modulus of elasticity and its importance in designing structural members*
- *Illustrate a stress–strain diagram and its properties*

© Springer International Publishing Switzerland 2015
P. Ghavami, *Mechanics of Materials*, DOI 10.1007/978-3-319-07572-3_6

6.1   Simple Stress

When a body is subjected to external forces a system of internal forces is developed. It is important in engineering mechanics to determine the intensity of these forces on the various cross section portions of the body, as the resistances to applied forces depend on these intensities. This intensity is called *stress* and it is a measure of the resisting forces. Stress is determined by dividing the total applied load (force) F, by the total area of loaded cross section A. This is expressed as

$$\sigma = F/A$$

The unit of stress in the metric system is newtons per square meter (N/m^2). One newton per square meter is equivalent to one pascal (Pa). The unit of stress in the English system is pounds per square inch (psi).

There is, however, another unit in engineering calculations: kilo-pounds, which is equivalent to 1,000 lb and is abbreviated as kip. In this case, the unit of stress will be kilo-pounds per square inch, or ksi.

6.2   Components of Stress

The internal resistance F is decomposed into component F_n perpendicular to the plane (known as *normal force*) and a force component F_t, parallel to the plane (known as *shear force*). The quantities F_n/A and F_t/A represent average normal and shear forces per unit area. These quantities are called *normal stress* and *shear stress* respectively. In general, normal and shear stresses are not uniformly distributed over the entire area of interest, and their magnitudes shall be found at any point within the area. However, in special cases where the components of F_n and F_t are uniformly distributed over the entire area A, the above equations for normal and shear stresses can be applied.

6.3   Tensile and Compressive Stresses

Tensile and compressive stresses are normal stresses developed within a body as a result of external forces. These stresses are axial stresses and act along the longitudinal axis of the member. Their action is in the same direction as the applied force.

In the tensile or compressive stress, the force is applied normal to the cross section under consideration, however if the transverse force is applied to a member the internal forces develop in the plane of the cross section and they are called *shearing forces* (Fig. 6.1). Since shear distribution in the cross section of the member is not necessarily uniform, an average shear stress is used. Examples of shear stress are found in bolts, rivets and pins that are being used for connecting of various structural members.

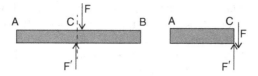

Fig. 6.1

The relationship between average shear stress in the cross section and the shear (F) can be expressed as:

$$\tau_{ave} = F/A$$

Figure 6.2, shows the rivet connection under tension force F and the shear stress will be developed in the rivet section connecting the plates. In this case, the force F is considered as the shear in the rivet section.

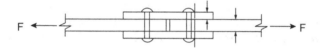

Fig. 6.2

6.4   Definitions

The following are the useful terms that are extensively used in engineering generally, and mechanics of materials especially.

Allowable Stress	*Allowable stress* is the maximum stress that a member can safely sustain in service. It is also called *design stress*.
Brittleness	*Brittleness* is the property of a material to fail without excessive deformation
Ductility	*Ductility* is the property of a material that allows the material to elongate under certain stress without breaking. Some materials like copper, and gold are relatively more ductile than others.
Elasticity	*Elasticity* is the ability of a material to return to its size and shape after elastic deformation.
Elastic Limit	*Elastic limit* is the highest value of the stress for which the material behaves elastically.
Hardness	*Hardness* is the resistance of a material to penetration and abrasion.
Malleability	*Malleability* is the property of a material with which it deforms under compressive stress without rupture. It is the tendency that the material to be hammered or rolled into sheets.

Plasticity	*Plasticity* is the property of a material with which it deforms under stress without rupture and not to return to its original size and shape.
Stiffness	*Stiffness* is the resistance of material to bending or deformation.
Strength	*Strength* of a material is the ability to withstand high forces without failure. Some materials represent better resistance than others. For instance, compressive and tensile strengths of steel are the same, whereas concrete is strong in compression and weak in tension.
Toughness	*Toughness* is the resistance of material to high forces without breaking.

6.5   Normal Stresses Under Axial Tension or Compression

As we discussed earlier, tensile and compressive internal forces develop in the structural members when they are under the axial loads. The internal forces will distribute over the entire cross section and create normal stresses in the cross section of the member. Figure 6.3 shows the force distribution on the cross section of the bar directed to the left when the force F is directed to the right.

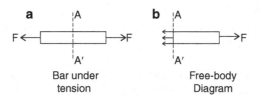

Fig. 6.3

When the member is under *tension*, the stress is called *tensile stress*. If the member is under *compression*, then the stress is called *compressive stress*. Tensile and compressive stresses are called normal stresses, since the force F and the force distribution are both perpendicular to the cross section of the member (Fig. 6.4). Normal stress formula $\sigma = F/A$ discussed earlier, can be applied for tensile or compressive stresses conditionally the force F is centrally applied to the cross section of the member.

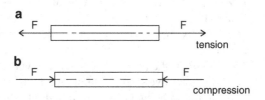

Fig. 6.4

Note that the sense of the stress is also important. It cannot be determined from the sign of the force vector. If the stress is tending to stretch the body or pull it apart, it is called tension. Any tension is considered positive. If the stress is compressive or squashing the body, it is called compression and carries a negative sign.

Example 6.1 Tensile forces equal to 5,000 N, are applied to the both sides of bar shown (Fig. 6.5). Calculate the tensile stress on the cross section of the bar with the dimensions of 5×5 mm.

Fig. 6.5

Solution

$$A = 0.005\,\text{m} \times 0.005\,\text{m} = 2.5 \times 10^{-5}\,\text{m}^2$$

$$\sigma = F/A = 5,000\,\text{N}/2.5 \times 10^{-5}\,\text{m}^2 = 2 \times 10^8\,\text{N}/\text{m}^2 \quad \text{or} \quad 200\,\text{MPa}$$

Example 6.2 Calculate the tensile stress in a wire 0.2 in. in diameter shown (Fig. 6.6) when it is subjected to an axial force of 250 lb.

Fig. 6.6

Solution

$$A = \pi(0.2\,\text{in.})^2/4 = 0.0314\,\text{in.}^2$$

$$\sigma = F/A = 250\,\text{lb}/0.0314\,\text{in.}^2 = 7,962\,\text{psi}$$

Example 6.3 Find the tensile stress in each of the solid cylinders shown (Fig. 6.7) if a mass of 5,000 kg is applied.

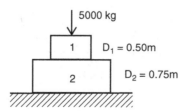

Fig. 6.7

Solution

$$\text{Load} = \text{mass} \times \text{acceleration due to gravity}$$
$$= (5{,}000\,\text{kg})\,(9.8\,\text{m/s}^2) = 49{,}000\,\text{N}$$

$$A_1 = \pi(0.5\,\text{m})^2/4 = 0.196\,\text{m}^2$$

$$A_2 = \pi(0.75\,\text{m})^2/4 = 0.442\,\text{m}^2$$

$$\sigma_1 = F/A_1 = 49{,}000\,\text{N}/0.196\,\text{m}^2 = 2.5 \times 10^5\,\text{Pa}$$

$$\sigma_2 = F/A_2 = 49{,}000\,\text{N}/0.442\,\text{m}^2 = 1.1 \times 10^5\,\text{Pa}$$

Example 6.4 Calculate the tensile stress in a steel bar 1.5 in. × 1.5 in. cross section shown (Fig. 6.8) if it is subjected to an axial load of 110 kips.

Fig. 6.8

Solution

$$A = 1.5\,\text{in.} \times 1.5\,\text{in.} = 2.25\,\text{in.}^2$$

$$\sigma = F/A = 110\,\text{kips}/2.25\,\text{in.}^2 = 48.89\,\text{ksi} \quad \text{or} \quad 48{,}890\,\text{psi}$$

Example 6.5 A circular tube with an outside diameter of 40 mm and inside diameter of 20 mm shown (Fig. 6.9) is under a compressive force of 50,000 N. Determine the compressive stress in the tube.

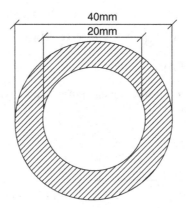

Fig. 6.9

Solution The effective area of cross section is the difference between outside circle area and the inside circle area.

$$A_{effective} = \pi D_o^2/4 - \pi D_i^2/4 = \pi/4 \times (D_o^2 - D_i^2)$$
$$= \pi/4 \times (0.0016\,\mathrm{m}^2 - 0.0004) = 9.4 \times 10^{-4}\,\mathrm{m}^2$$

$$\sigma = 50,000\,\mathrm{N}/9.4 \times 10^{-4}\,\mathrm{m}^2 = 5.31 \times 10^7\,\mathrm{Pa}$$

6.6 Shear Stress

In Sect. 6.3 we discussed tensile and compressive stresses and we called them *normal stresses,* because they act perpendicularly to the surface. *Shear stresses,* however, are developed in a parallel direction, or tangentially to the surface (Fig. 6.10).

Fig. 6.10

Example 6.6 Figure 6.11 shows two steel plates connected using four ½-in. diameter bolts. If the tensile load is 20,000 lb, find the average shear stress in the bolts.

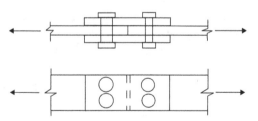

Fig. 6.11

Solution Total area of bolts $= (\pi D^2/4) \times 4 = \pi D^2 = \pi(0.5\,\mathrm{in.})^2 = 0.785\,\mathrm{in.}^2$
Total shear stress on the bolts is

$$\tau_{ave} = F/A$$
$$= 20,000\,\mathrm{lb}/0.785\,\mathrm{in.}^2 = 2.55 \times 10^4\,\mathrm{psi}$$

Example 6.7 A 400-lb tensile load is carried by a 1.00-in. diameter rivet as shown in Fig. 6.12. Determine the shear stress in the rivet.

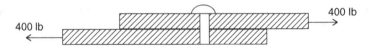

Fig. 6.12

Solution

$$A = \pi D^2/4 = \pi(1)^2/4 = 0.785 \text{ in.}^2$$

$$\tau = F/A = 400 \text{ lb}/0.785 \text{ in.}^2 = 509.3 \text{ psi}$$

Example 6.8 Calculate the shearing stress in the pins with diameter of ½-in. shown in Fig. 6.13. A tensile load of 250 lb is applied.

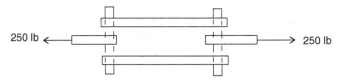

Fig. 6.13

Solution

$$A = 2 \times \left(\pi D^2/4\right) = 2 \times \left[\pi(1/2)^2/4\right] = 0.393 \text{ in.}^2$$

$$\tau = F/A = 250 \text{ lb}/0.393 \text{ in.}^2 = 636.6 \text{ psi}$$

Example 6.9 Calculate the shearing stress in the bolts with the diameter of ¼ in. for connecting the plates shown (Fig. 6.14). A tensile load of 10,000 lb is applied.

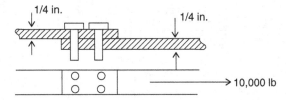

Fig. 6.14

Solution

$$A = 4 \times \left(\pi D^2/4\right) = 4 \times (\pi)(0.25 \text{ in.})^2/4 = 0.196 \text{ in.}^2$$

$$\tau = F/A = 10,000 \text{ lb}/0.196 \text{ in.}^2 = 5.09 \times 10^4 \text{ psi}$$

Shear stress can also be calculated when a force F is applied to punch a hole through a plate.

In this case, the shear stress is resisted by a cylindrical area with a diameter equal to the punch diameter and a height equal to the thickness of the plate.

Example 6.10 A punching machine punches a 1-in. diameter hole through a ¼-in. plate. Calculate the force needed to punch through the plate. Assume the shear stress is 25,000 psi.

Solution
$$A = (\pi)(d)\,(\text{plate thickness})$$

$$A = (\pi)(1.00\,\text{in.})(0.25) = 0.785\,\text{in.}^2$$

$$F = \tau \cdot A = (25,000\,\text{psi})(0.785\,\text{in.}^2) = 1.96 \times 10^4\,\text{lb}$$

6.7   Allowable Stress

If a member is loaded beyond its ultimate stress, it will fail or rupture. In engineering structures, it is essential that the structure not fail. Thus, the design is based on some lower value called *allowable stress* or *design stress*. If, for example, a certain type of steel is known to have an ultimate strength of 110,000 psi, a lower allowable stress would be used for design, say 55,000 psi. This allowable stress would allow only half the load the ultimate stress would allow. Allowable stress values are different for different materials, and they are tabulated and recommended by the International Building Code Association.

The ratio of the ultimate stress to the allowable stress is known as the *factor of safety*.

$$\text{factor of safety} = \text{ultimate strength} \,/\, \text{allowable stress}$$

Example 6.11 Determine the required size for a steel rod to support a tensile load of 50,000 lb if the allowable tensile stress of the steel is 25, 000 psi.

Solution
$$\sigma = F/A, \quad \text{or} \quad A = F/\sigma = 50,000\,\text{lb}/25,000\,\text{psi} = 2\,\text{in.}^2$$

$$A = \pi D^2/4 = 2\,\text{in.}^2$$

Solving for D, we have:

$$D = 1.6\,\text{in.}$$

Example 6.12 Compute the required dimension d of a steel column subjected to an axial compressive load of 3×10^6 N (Fig. 6.15). The allowable compressive stress is 8×10^7 Pa.

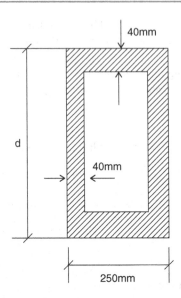

40mm

40mm

d

250mm

Fig. 6.15

Solution

$$A_{\text{eff}} = F/\sigma = 3 \times 16^6 \, \text{N}/8 \times 10^7 \, \text{Pa} = 0.0375 \, \text{m}^2 = 3.75 \times 10^4 \, \text{mm}^2$$

$$170(d - 80) = 170d - 13,600 \, \text{mm}^2 \ \text{area of inside rectangle}$$

$$A_{\text{eff}} = 250d - (170d - 13,600) = 250d - 170d + 13,600$$
$$= 80d + 13,600$$

$$3.75 \times 10^4 \, \text{mm}^2 = 80d + 13,600$$

$$d = 23,900 \, \text{mm}^2/80 \, \text{mm} = 298.8 \, \text{mm}$$

Practice Problems

1. Calculate the tensile stress in a wire 4 mm in diameter when it is subjected to an axial tensile force of 2,000 N.
2. Calculate the tensile stress in a metal rod with a cross section 0.5 in. by 2.5 in. if it is under a tensile force of 1,500 lb.
3. Calculate the compressive stress in a circular bar 4 in. in diameter when it is subjected to an axial compressive force of 50,000 lb.
4. Calculate the axial compressive force on the frame cross section used in the machine shown in Fig. 6.16 if the compressive stress is 50 MPa.

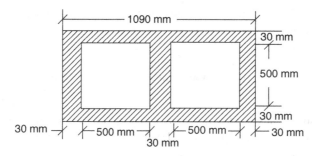

Fig. 6.16

5. Calculate the allowable axial tensile load for a steel bar with a cross section 20 mm by 80 mm if an allowable tensile stress of 2×10^8 Pa developed in the bar.
6. A punching machine punches a 0.5-in. diameter hole through a 3/16-in. plate.
     If the applied force to punch the hole through the plate is 35,000 lb, calculate the shear stress developed in the metal plate.
7. Calculate the maximum axial tensile load that can be hung from a steel wire ½ in. in diameter if the allowable axial tensile stress for the steel is 25,000 psi.

6.8   Tensile and Compressive Strain

When a deformable body is subjected to stresses, it undergoes deformation. Deformation is normally accompanied by a geometrical change of the body. In other words, the dimensions of a body may be *elongated* or *shortened*. The total deformation, or change in the shape or size of a body under the force is called elongation, shown by symbol δ. If, for example, the original length of a rod is L, and after the rod experiences a tensile force is L_f, the elongation is

$$\delta = L - L_f$$

In most engineering problems, it will be more meaningful to express the deformation on a unit basis. Therefore, we use tensile or compressive strain. Strain, represented by the symbol ε, is defined as the total deformation (δ) of a body divided by the original length of the body to indicate its unit deformation and is written as

$$\varepsilon = \delta/L$$

Note that, by definition, strain is a dimensionless quantity. However, it is common to express strain in units of length divided by units of original length, such as inches per in or mm/mm.

Example 6.13 Calculate the total elongation for a certain material 20 m in length if the strain is 0.00045 mm/mm.

$$\begin{aligned}\delta &= \varepsilon \times L \\ &= (0.00045\,\text{mm/mm})(20,000\,\text{mm}) = 9\,\text{mm}\end{aligned}$$

Strain is sometimes used as a synonym for deformation. For example, if the elongation of a 20-ft rod under a tensile force is 0.15 in. (i.e. $\delta = 0.15$ in.), the strain or deformation is

$$\varepsilon = \delta/L = 0.15\,\text{in.}/(20 \times 12\,\text{in.}) = 0.000625\,\text{in./in.}$$

In this example, the total elongation is 0.15 in. and the deformation (strain) is 0.000625 and has no *physical unit*, so the strain would be the same whether we are working with English units or metric units. Figures 6.17 and 6.18 show the applied force and deformation δ for tensile force and compressive force.

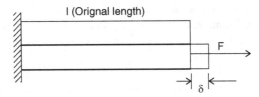

Fig. 6.17

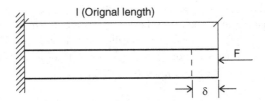

Fig. 6.18

Example 6.14 A certain metal rod 10.00 m long (unstretched length) is stretched to 10.01 m. Find the strain.

Solution Deformation or total elongation can be calculated from the previous formula

$$\delta = 10.01 - 10.00 = 0.01\,\text{m}$$

and the strain is

$$\varepsilon = \delta/L = 0.01\,\text{m}/10\,\text{m} = 0.001 \ (\text{No physical unit!})$$

Example 6.15 A 3-m rod is suspended from the ceiling. If the deformation is 2 mm, find the axial strain in the rod.

Solution

$$\begin{aligned} \varepsilon &= \delta/L \\ &= 2\,\text{mm}/3,000\,\text{mm} \\ \varepsilon &= 0.000667 \end{aligned}$$

Practice Problems

1. A telephone wire 100 m long is stretched by a tensile force. Find the tensile strain if the final length is 100.15 m.
2. A bar 0.500 m long undergoes a deformation of 0.00022 m. Find the strain.
3. Given $\delta = 1.5$ in. and $L = 50$ ft, calculate ε.
4. Given $\delta = 0.7$ ft and $\varepsilon = 0.00020$, calculate L.
5. A column 10 ft long is compressed 1/17 in. Find the strain.
6. A metal bar 0.2 m long is stretched 0.053 mm. Find the strain.

6.9   The Relationship Between Stress and Strain

The relationship between stress and strain exists in most engineering testing of materials. Experiments show that when a specimen is subjected to tensile stress, there is a closely proportional increase in strain, such that strain increases in direct relation to stress as long as stress is kept within certain limits. If the generated stress exceeds the limiting value, the corresponding strain will no longer be proportional to the stress. This limiting value is known as the *proportional limit.*

Robert Hooke (1635–1703), discovered that stress and strain are directly proportional to each other. He plotted the calculated values of stress and strain for increasing the load and called it a stress–strain diagram, known as *Hooke's law.* A stress–strain diagram shows a linear relationship between stress and strain as long as the values of stress are not too high (Fig. 6.19). Thomas Young (1773–1829) found out that the ratio of the stress and strain values remained constant. This constant is now called *Young's modulus* or the *modulus of elasticity* (*E*). Note that the slope of the straight line part of the diagram is known as the modulus of elasticity of the material.

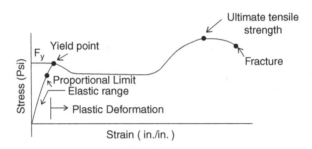

Fig. 6.19 Stress-strain diagram

The mathematical relation between stress, strain, and the modulus of elasticity is expressed by the equation

$$E = \text{stress/strain.} = \sigma/\varepsilon$$

Modulus of elasticity E is an important property in engineering with which engineers can predict the behavior of materials. It can be interpreted as the measure of the stiffness of a material. That means that materials with higher values of modulus of elasticity are stiffer than others, and they have more ability to resist a deformation. The value of the modulus of elasticity is constant for a specific material and can be found in a table for a number of common materials. The unit for modulus of elasticity in the metric system is expressed in pascals (Pa) or megapascals (MPa) and, in the English system is pounds per square inch (psi).

Example 6.16   A steel rod 1 in. in diameter and 10 in. in length stretches 0.009 in. when subjected to a tensile load of 16,000 lb. Find the modulus of elasticity.

Solution
$$A = \pi D^2/4 = (\pi/4)(1\,\text{in.})^2 = 0.7854\,\text{in.}^2$$

$$\sigma = F/A = 16,000\,\text{lb}/0.7854\,\text{in.}^2 = 2.04 \times 10^4\,\text{psi}$$

$$\varepsilon = \delta/L = 0.009\,\text{in.}/10\,\text{in.} = 0.0009\,\text{in.}/\text{in.}$$

and

$$\begin{aligned} E &= \text{stress/strain.} = \sigma/\varepsilon \\ &= 2.04 \times 10^4\,\text{psi}/0.0009\,\text{in.}/\text{in.} \\ E &= 2.3 \times 10^7\,\text{psi} \end{aligned}$$

Example 6.17   A 5,000-N weight is hanging from a cable 3 m long. Find the diameter of the cable if the deformation is 5 mm and $E = 200,000$ MPa.

Solution

$$\varepsilon = \delta/L = 5\,\text{mm}/3,000\,\text{mm} = 0.001667\,\text{mm}/\text{mm}$$

$$E = \sigma/\varepsilon$$

$$\sigma = E \cdot \varepsilon = \left(200,000 \times 10^6\,\text{N}/\text{m}^2\right)(0.001667) = 3.33 \times 10^8\,\text{N}/\text{m}^2$$

$$A = F/\sigma = 5,000\,\text{N}/3.33 \times 10^8\,\text{N}/\text{m}^2 = 1.5 \times 10^{-5}\,\text{m}^2$$

$$1.5 \times 10^{-5}\,\text{m}^2 = \pi/4\left(d^2\right)$$

$$d = 4.4 \times 10^{-3}\,\text{m} \quad \text{or} \quad d = 4.4\,\text{mm}$$

Example 6.18 Find the amount of the strain in a copper bar that is subjected to a compressive stress of 10,000 psi. $E_{\text{copper}} = 1.7 \times 10^7$ psi.

Solution

$$\varepsilon = \sigma/E = 10,000\,\text{psi}/1.7 \times 10^7\,\text{psi} = 0.0006$$

Example 6.19 A steel bar ½ in. in diameter and 10 in. long is subjected to a tensile load of 5,000 lb. How much does it stretch? $E_{\text{steel}} = 2.0 \times 10^7$ psi.

Solution

$$A = \pi/4\left(D^2\right) = \pi/4(0.5\,\text{in.})^2 = 0.196\,\text{in.}^2$$

$$\sigma = F/A = 5,000\,\text{lb}/0.196\,\text{in.}^2 = 2.55 \times 10^4\,\text{psi}$$

$$\varepsilon = \sigma/E = 2.55 \times 10^4\,\text{psi}/2.0 \times 10^7\,\text{psi} = 0.00127$$

$$\delta = \varepsilon \cdot L = 0.00127 \times 10\,\text{in.} = 0.0127\,\text{in.}$$

6.10 More Useful Formulas

A more useful formula can be derived using the previous equations.

$$E = \sigma/\varepsilon = (F/A)/(\delta/L)$$

then

$$E = FL/A\delta$$

or

$$\delta = FL/AE$$

We can see that the above formulas could be very useful in calculations and also that it takes a shorter time to get the problem done.

Example 6.20 A rectangular aluminum bar 600 mm long with a cross sectional area of 20×30 mm is subjected to an axial tensile load of 20,000 N. Find the total axial deformation. $E_{\text{aluminum}} = 7 \times 10^4$ MPa.

Solution Given:

$$
\begin{aligned}
A &= 20 \times 30\,\text{mm} = 600\,\text{mm}^2 = 600 \times 10^{-6}\,\text{m}^2 \\
L &= 600\,\text{mm} = 0.6\,\text{m} \\
F &= 20,000\,\text{N} \\
E &= 7 \times 10^4\,\text{MPa} = 7 \times 10^{10}\,\text{N/m}^2
\end{aligned}
$$

$$
\begin{aligned}
\delta &= FL/AE = 20,000\,\text{N} \times 0.6\,\text{m} / \left(600 \times 10^{-6}\,\text{m}^2\right)\left(7 \times 10^{10}\,\text{N/m}^2\right) \\
\delta &= 2.86 \times 10^{-4}\,\text{m} = 0.286\,\text{mm}
\end{aligned}
$$

Practice Problems

1. A hard plastic bar 20 in. long with a 0.3-in. diameter is subjected to a tensile load of 1,200 lb. Find the total deformation. Assume that $E = 5 \times 10^5$ psi.
2. A steel wire 30 m long stretches 52 mm when subjected to a tensile load of 5,000 N. Find the diameter of the wire. Assume that $E_{\text{steel}} = 2 \times 10^{11}$ Pa.
3. A 20-mm diameter aluminum rod 2 m long is subjected to an axial tensile load of 55,000 N. Find the total deformation. Assume that $E_{\text{aluminum}} = 7 \times 10^4$ MPa.
4. A cast iron block 200 mm long with a cross sectional area of 100 mm $\times$ 250 mm is subjected to an axial compressive load. Find the compressive force if the deformation is 0.0427 mm. Assume that $E_{\text{cast iron}} = 165$ GPa.
5. A steel bar 5 ft long with a cross sectional area of 1.5 in.^{2} is subjected to an axial tensile load of 5,000 lb. Calculate the total deformation. Assume that $E = 3 \times 10^7$ psi.
6. An aluminum wire with a 3-mm diameter is subjected to an axial tensile load of 550 N and stretches by 20 mm. Assuming that $E = 6.9 \times 10^{10}$ Pa, find the length of the wire.
7. A wire 10 ft long and 0.5 in.^{2} in cross sectional area is subjected to an axial tensile load of 5,000 lb and stretches by 0.08 ft. Calculate the modulus of elasticity.
8. A metal wire 3 m long and 0.75 mm in diameter is subjected to a 300-kg mass and stretches by 15 mm. Find: (a) the stress; (b) the strain; and (c) the modulus of elasticity.

Chapter Summary

Engineers must first identify the external forces acting on a structural member, and then calculate the internal resistance, called *stress*, developed by the member. Based on this, and other related properties such as *strain* and the *modulus of elasticity*, the designer is able to select the material and design the size and shape of the member that will properly resist the applied forces.

The materials used vary in quality and physical and mechanical characteristics. Hence, it is important that a designer have good knowledge of the properties of the materials being used. In this chapter, we introduced the concepts of *stress*, *strain*, and *deformation*, and the relationship between stress and strain in different materials.

Stress is defined as the force per unit area for a specific material under loading. The unit of stress in the English system is psi, and in the metric system is Pa, or MPa. There are three *normal stresses*: *tensile stress, compressive stress, and shear stress*. These may be calculated using the stress formula:

$$\sigma = F/A$$

Strain is defined as the total deformation, or stretching, per unit length of the member. The formula to calculate the strain is

$$\varepsilon = \delta/L$$

Stress is directly proportional to the strain:

$$\sigma = E \cdot \varepsilon$$

This relation is known as *Hooke's law* and the proportionality constant is called the *modulus of elasticity* (E) of the material. The modulus of elasticity is a measure of a material's resistance to deformation. The above linear equation is valid provided that the stress does not exceed the proportional limit of the material. The proportional limit is the maximum stress for which stress is proportional to strain.

Review Questions

1. What is stress?
2. What is strain?
3. What are the units of stress and strain?
4. What is a tensile load?
5. What is a compressive load?
6. What is shear?
7. What is the relation between stress and strain?
8. What is Hooke's diagram, and where is it valid?
9. What is the modulus of elasticity, and what is meant by it in engineering problems?
10. What is the allowable stress?
11. What is the ultimate stress?
12. What is the relation between allowable stress and ultimate stress?
13. What is the proportional limit?

Problems

1. A tie rod 10.0 ft long and 2.2 in. in diameter is subjected to a tensile force of 3,000 lb. Find the stress, strain, and the total deformation. Use $E = 2.8 \times 10^7$ psi.

2. What is the tensile strain in a specimen of material if it is subjected to a tensile force of 2×10^8 N/m^2? Use $E = 1.1 \times 10^{11}$ Pa.

3. A telephone wire 100 m long and 2.5 mm in diameter is subjected to a tensile force of 400 N. Find the stress, strain, and the modulus of elasticity. Assume that the final length is 100.30 m after stretching.

4. A 200–mm long metal alloy tube of 40 mm outer diameter and 30 mm inner diameter is subjected to a compressive load of 30, 000 N. Find the stress, strain, and final length. Use $E = 9 \times 10^{10}$ Pa.

5. Find the magnitude of the tensile force acting on a steel bar 1.50 in. in diameter if the strain in the bar is 0.0015. Use $E = 3 \times 10^7$ psi.

6. A rectangular steel bar ¾ in. × 7/8 in. and 20 in. long is subjected to a tensile load of 3,000 lb. Find the stress, strain, and the total deformation. Use $E = 3 \times 10^7$ psi.

7. A steel cable 50 ft long and 0.5 in. in diameter is subjected to a tensile force and stretches 0.37 in. Find the stress. Use $E = 3 \times 10^7$ psi.

8. A copper wire 5 m long and 2 mm in. diameter is subjected to a tensile load of 600 N, and stretches 12.5 mm. Calculate the modulus of elasticity of the wire.

9. A steel circular cross section of a column is subjected to a tensile load of 500,000 N.
   Find the size of the cross section of the column if the allowable stress is 1×10^8 Pa.

10. A steel circular cross section with a diameter of 2 in. is subjected to an axial tensile force. Find the force if the resulting tensile strain is 0.0005.

Stephen P. Timoshenko (1878–1972)

Stepan Prokopovych Timoshenko was born on December 22, 1878 in the village of Shpotivka in the Ukrain. Timoshenko's early life seems to have been a happy one in pleasant rural surroundings. He studied at a "realnaya" school from 1889 to 1896. Timoshenko continued his education towards a university degree at the St. Petersburg Institute of Engineering. After graduating in 1901, he stayed on teaching in this same institution from 1901 to 1903 and then worked at the St. Petersburg Polytechnic Institute under Viktor Kyrpychov 1903–1906. His restlessness and discontent with the educational system extant in Russia at that time motivated the young Timoshenko to venture out to explore, examine, and assimilate diverse pedagogical views and cultures in France, Germany, and England. In 1905 he was sent for 1 year to the University of Göttingen where he worked under Ludwig Prandtl.

In the fall of 1906 he was appointed to the Chair of Strengths of Materials at the Kyiv Polytechnic Institute. Thanks to his tormented spirit at this institute, Timoshenko took the plunge to writing his maiden Russian classic, *Strength of Materials* in 1908 (Part I) and 1910 (Part II). From 1907 to 1911 as a professor at the Polytechnic Institute he did research in the area of finite element methods of elastic calculations, and did excellent research work on buckling. He was elected dean of the Division of Structural Engineering in 1909.

In 1911 he was awarded the D. I. Zhuravski prize of St. Petersburg; he went there to work as a Professor in the Electro-technical Institute and the St. Petersburg Institute of the Railways (1911–1917). During that time he developed the theory of elasticity and the theory of beam deflection, and continued to study buckling.

In 1922 Timoshenko moved to the United States where he worked for the Westinghouse Electric Corporation from 1923 to 1927, after which he became a faculty professor at the University of Michigan where he created the first bachelor's and doctoral programs in engineering mechanics. His textbooks have been published in 36 languages. His first textbooks and papers were written in Russian; later in his life, he published mostly in English.

The following 3 years 1935–1937, Timoshenko teamed up with Gere and Young for three more books: a condensed guide of strength of materials, elastic stability, and engineering mechanics. These unique texts explore intricate mathematical techniques to explain some subtle aspects underlying elasticity and stability to give new insight into the behavior of solids and structures for engineering design. From 1936 onward he was a professor at Stanford University.

The year 1953 saw the great Timoshenko epic, *The History of Strength of Materials* with a brief account of the history of the theory of elasticity and structural mechanics. Tracing the history all the way back to Archimedes, he carries the reader through the period of Leonardo da Vinci, Galileo, Hooke, Newton, Mariotte, Bernoulli, Euler, Lagrange, and Coloumb, reaching the end of the eighteenth century.

This missionary zeal of Timoshenko for writing books for improving teaching and for guiding practical engineers has played a key role in uplifting technical education worldwide, but more emphatically in the United States. In 1960 he moved to Wuppertal (Western Germany) to be with his daughter. He died in 1972 and his ashes are buried in Alta Mesa Memorial Park, Palo Alto, California.

Torsion in Circular Sections

7

Overview

In the previous chapter, we discussed the fundamental concepts underlying normal stress, strain, and shear stress in structural members. These stresses are developed in a body to resist external forces. In this chapter we will study the shear stress in a circular cross section when it is subjected to torsion. We will learn the concept of torque, and the twisting action applied to the axis of a member, and then we will determine the shear stresses developed to resist the application of torque, or twisting moment.

Learning Objectives

Upon completion of this chapter, you will be able to calculate the torque, or twisting moment, on a circular section of a shaft and to calculate the torsional shearing stresses developed on the cross-sectional planes of a shaft. Also, you will learn how to calculate the transmitting power (in horsepower) developed by a rotational shaft. Your knowledge, application, and problem solving skills will be determined by your performance on the chapter test.

7.1　Torque

When a circular bar is subjected to a twisting action it tends to twist; the stresses developed in the bar are called torsional stresses. In other words, the bar is subjected to torque, or twisting moment, and is said to be in torsion, or under torsional load. Torque is expressed in units of length and force. The unit of torque in the English system is inch-pounds, and in the metric system is newton-meters. The torque on a shaft is applied through pulleys or gears at different locations of the shaft, and the amount of the torque at any point on the shaft is the moment of the force along the axis of the shaft (Fig. 7.1).

© Springer International Publishing Switzerland 2015
P. Ghavami, *Mechanics of Materials*, DOI 10.1007/978-3-319-07572-3_7

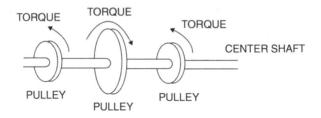

Fig. 7.1 Torsion on a drive shaft

Example 7.1 Determine the torque on the shaft shown in Fig. 7.2 at point A.

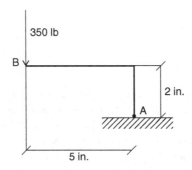

Fig. 7.2

Solution A force of 350 lb is applied to the left of point A. Therefore, to calculate the torque at point A with respect to the axis of rotation, we simply multiply the force by the moment arm of 5 in.

$$T = -350 \, \text{lb} \times 5 \, \text{in.} = -1{,}750 \, \text{in.-lb}$$

The negative sign means that the torque is counterclockwise.

Example 7.2 Figure 7.3 depicts the shaft and pulleys on a power train. Determine the torque at section R-R.

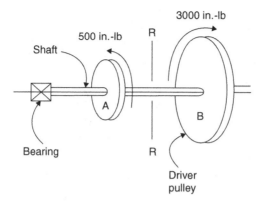

Fig. 7.3

Solution   The only externally applied torque to the left of section R-R is -500 in.-lb acting on pulley A.

The applied torque to the right of section R-R is 3,000 in.-lb. Therefore, the total externally applied torque is

$$\sum T = -500 \,\text{in.-lb} + 3{,}000 \,\text{in.-lb} = 2{,}500 \,\text{in.-lb}$$

For equilibrium, the above torque (twisting moment) must equal the resisting moment.

7.2   Torsional Shearing Stress in a Solid Circular Shaft

Let's take a look at a circular shaft (Fig. 7.4) that is fixed at one end and free at the other end. When the shaft is subjected to torque, the shearing stresses occur in the circular cross section of the shaft. The direction of the shearing stress at any point in the cross section is perpendicular to the radius of the shaft at the point of interest (Fig. 7.5).

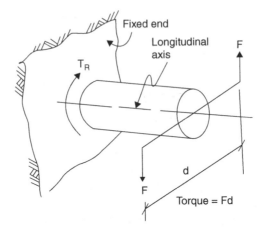

Fig. 7.4

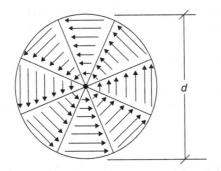

Fig. 7.5

Consider an infinitesimal area a located on the shaft. It can be shown that the shear stress on this cross section at a distance r from the center is proportional to the maximum shear stress developed at the outer surface with a distance c from the center ($c = d/2$). This is expressed as

$$\tau = \tau_{max}(r/c) \tag{7.1}$$

where

τ = shear stress on cross section area a
τ_{max} = max. shear stress at outer surface of the shaft
r = distance from cross section area a to the center of shaft
c = distance from the outer surface to the center of shaft ($c = d/2$)

From Chap. 6, we know

$$F = \tau \times A = \tau_{max}(r/c)A \qquad (7.2)$$

where F is shear force over the shaft.
The resisting internal torque is calculated as

$$T = F \times r \qquad (7.3)$$

Substituting from Eq. (7.2) into Eq. (7.3), we obtain an equation for calculating the maximum shear stress developed on the cross section of the shaft.

$$T = \tau_{max}(r/c)A \cdot r = \tau_{max}(r^2/c)A \qquad (7.4)$$

Total torque for entire area,

$$T = \sum (\tau_{max}/c)(Ar^2) \qquad (7.5)$$

where $Ar^2 = $ polar moment of inertia
Solving for the shear stress (limiting case),

$$\tau_{max} = T \cdot c/J \qquad (7.6)$$

For a solid circular shaft, $(c = d/2)$, $J = \pi\, d^2/16$, and Eq. 7.6 becomes

$$\tau_{max} = 16T/\pi d^3 \qquad (7.7)$$

where τ_{max} is the shearing stress in psi. T is the torque in in.-lb, and d is the diameter.

These units are all in the English system; in the metric system, we must use the metric units explained earlier, but the formula is the same.

Limitations
1. For design purposes, the above formula (7.6) is only valid for a solid circular shaft; for hollow shafts, J must be calculated for inner and outer diameters.
2. Developed shear stresses are assumed to be below the proportional limit, and shaft materials follow Hooke's law. Beyond the elastic limit the formula will no longer be valid.

Example 7.3 A 1,700-lb load is acting on a pulley of diameter 3 ft that is attached to a solid shaft. Calculate the maximum shearing stress in the shaft. Use the diameter of the shaft $d = 4$ in.

Solution
$$T = F \cdot r = 1,700\,\text{lb} \times (3\,\text{ft} \times 12/2) = 30,600\,\text{in.-lb}$$

$$d = 4\,\text{in.}$$

$$\tau_{max} = 16T/\pi d^3$$
$$= 16(30,600)/(\pi)(4\,\text{in.})^3$$
$$\tau_{max} = 2,435\,\text{psi}$$

Example 7.4 A 20,000-lb load is applied to a pulley 26 in. in diameter on a solid shaft. Calculate the diameter of the shaft if the maximum shearing stress is 10,000 psi.

Solution

$$T = F \cdot r = 20,000\,\text{lb} \times 26/2 = 260,00\,\text{in.-lb}$$

$$d = ?$$

$$\tau_{max} = 16T/\pi d^3$$

$$d = (16T/\pi \tau_{max})^{1/3}$$

$$d = [(16)(260,000\,\text{in.-lb})/(\pi)(10,000\,\text{psi})]^{1/3}$$

$$d = 5\,\text{in.}$$

Example 7.5 Calculate the torque that must be applied to a solid circular shaft 3 in. in diameter to develop a maximum shearing stress of 15,000 psi.

Solution

$$\tau_{max} = 16T/\pi d^3$$

or

$$T = \pi d^3 \tau_{max}/16$$

$$T = (\pi)(3\,\text{in.})^3(15,000\,\text{psi})/16 = 79,521.6\,\text{in.-lb}$$

Example 7.6 Calculate the torque that must be applied to a solid circular shaft 12 mm in diameter to develop a maximum shearing stress of 50 MPa.

Solution

$$T = \pi d^3 \tau_{max}/16$$

$$T = (\pi)(0.012\,\text{m})^3(50 \times 10^6\,\text{N/m}^3)/16$$

$$T = 16.96\,\text{N.m}$$

Example 7.7 A 3,000 in.-lb twisting moment is applied to a solid shaft. Calculate the diameter of the shaft if the maximum shearing stress is 10,000 psi.

Solution

$$d = (16T/\pi\tau_{max})^{1/3}$$
$$= (16 \times 3,000\,\text{in.-lb.}/\pi \times 10,000\,\text{psi})^{1/3}$$
$$d = 1.15\,\text{in.}$$

7.3   Torsional Shearing Stress in a Hollow Circular Shaft

For a hollow circular shaft, we use outer diameter d_o, and inner diameter d_i (Fig. 7.6), and the maximum torsional shearing stress is

$$\tau_{max} = 16T\,d_o/\pi(d_o^4 - d_i^4)$$

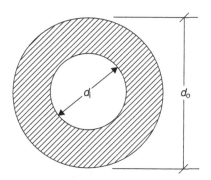

Fig. 7.6

Example 7.8 Calculate the torque that must be applied to a hollow circular shaft of 6 in. outer diameter and 5 in. inner diameter if the maximum torsional shearing stress is 15,000 psi.

Solution

$$\tau_{max} = 16T\,d_o/\pi(d_o^4 - d_i^4)$$

or

$$
\begin{aligned}
T &= \pi(d_o^4 - d_i^4)\tau_{max}/16\,d_o \\
&= \pi(d_o^4 - d_i^4)\tau_{max}/16\,d_o \\
&= \pi\left[(6)^4 - (5)^4\right](15,000)/16 \times 6 \\
T &= 329,376.35\,\text{in.-lb}
\end{aligned}
$$

Example 7.9 Calculate the maximum torsional shearing stress developed on a hollow circular shaft of 200 mm outer diameter and 150 mm inner diameter if the applied torque is 10,000 N.m.

$$\tau_{max} = 16T\,d_o/\pi\left(d_o{}^4 - d_i{}^4\right)$$
$$= 16(10,000)(0.2)/\pi\left[(0.2)^4 - (0.15)^4\right]$$
$$\tau_{max} = 9.3\,\text{Mpa}$$

Example 7.10 The socket shown (Fig. 7.7) is used to tighten the bolts with an applied torque.

Use the information shown to calculate the maximum torsional shearing stress developed on the socket. Use outer diameter of the socket $d_o = 1.00$ in. and inner diameter $d_i = 0.8$ in.

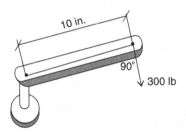

10 in.

90°

300 lb

Fig. 7.7

Solution Given:

$$d_o = 1.00\,\text{in.}$$
$$d_i = 0.8\,\text{in.}$$
$$L = 10\,\text{in.}$$
$$F = 300\,\text{lb}$$

$$T = F \times r = 300\,\text{lb} \times 10\,\text{in.} = 3,000\,\text{in.-lb}$$

$$\tau_{max} = 16T\,d_o/\pi\left(d_o{}^4 - d_i{}^4\right)$$
$$= 16(3,000)(1.00)/\pi\left[(1)^4 - (0.8)^4\right] = 25,879\,\text{psi}$$

Practice Problems

1. Calculate the torque given $F = 200$ N and $r = 50$ cm.
2. Calculate the torque given $F = 120$ lb and $r = 7$ in.
3. Calculate the maximum torsional shearing stress developed in a shaft 20 in. in diameter if the applied torque is 100,000 in.-lb.

4. Calculate the torque that must be applied to a solid circular shaft 3 in. in diameter if the maximum torsional shearing stress is 12,000 psi.
5. Calculate the torque that must be applied to a solid circular shaft 25 mm in diameter if the maximum torsional shearing stress is 55 MPa.
6. When 3,000 N.m of torque is applied to a solid circular shaft, it causes a maximum torsional shearing stress of 60 MPa. Calculate the diameter of the shaft.
7. When 4,000 N.m of torque is applied to the 85-mm outer diameter of a hollow circular shaft, it causes a maximum torsional stress of 65 MPa. Find the inner diameter of the shaft.
8. Calculate the torque that must be applied to a hollow circular shaft of 0.7 in. outer diameter and 0.4 in. inner diameter if the maximum torsional shearing stress is 5,000 psi.
9. A hollow circular shaft has a 70-mm outer diameter and a 20-mm inner diameter. A load of 15,000 N is applied to a 500-mm pulley on the shaft. Calculate the maximum torsional shearing stress in the shaft.
10. Calculate the maximum torsional shearing stress in a hollow circular shaft of 7 in. outer diameter and 4 in. inner diameter if the applied torque it needs to resist is 50,000 in.-lb.

7.4   Transmission of Power by a Rotating Shaft

A common objective of rotating shaft design is to transmit the power. When the torque is applied to a shaft and turns the shaft, work is performed. By definition, work is the product of the magnitude of the displacement and the component of the force in the direction of the displacement. This can be expressed as

$$W = F \cdot x$$

where x is the distance that a belt or cable moves on the pulley because of applied torque. Since the applied torque turns the shaft, this linear distance can be modified to circular distance. Therefore, the above formula for work can be written as

$$W = T\theta$$

where T is torque applied on the shaft, and θ is angle of rotation in radians. Power is defined as work done per unit time:

$$P = W/t$$

The unit of work in the metric system is N.m or Joule (J). The unit of time is the second and the unit of power is J/s = Watt (W). In the English system, the unit of work is the ft-lb; the unit of time is the second, and the unit of power is ft-lb/s. No special name is given to this unit of power. However, the following is the unit conversion between ft-lb/s and horsepower.

$$1 \text{ horsepower (hp)} = 550 \text{ ft lb/s} = 33,000 \text{ ft lb/min}$$

or

$$1 \text{ hp} = 39,600 \text{ in. lb/min}$$

The work done in one revolution is $2 \pi T$ and the new formula for work per minute is

$$W/\text{min} = 2 \pi T n$$

or

$$\text{Power} = \text{work/time}$$

T the unit of T is in.-lb, and n is the number of revolutions per min (RPM). Substituting this information in the formula for power (HP), we get:

$$\text{HP} = 2 \pi T n/396,000 = T n/63,025$$

Note that 39,600 in.-lb per minute is equal to 1 HP.

Example 7.11 A 2-in. diameter steel shaft is operating at 1,800 rpm. Calculate the maximum horsepower that can be transmitted by this shaft if the allowable shearing stress is 10,000 psi.

Solution We can calculate the maximum resisting torque from

$$
\begin{aligned}
T_{\text{max}} &= (\pi/16)d^3 \tau_{\text{all}} \\
&= (\pi/16)(2)^3(10,000) \\
&= 15,708 \text{ in.-lb.}
\end{aligned}
$$

$$
\begin{aligned}
\text{HP} &= T n/63,025 \\
&= (15,708)(1,800)/63,025 \\
&= 449 \text{ hp}
\end{aligned}
$$

Example 7.12 A 1.5-in. diameter steel shaft is operating at a certain rpm to transmit 15 hp. Calculate the rpm if the allowable shearing stress is 8,000 psi.

Solution

$$T_{max} = (\pi/16)d^3\tau_{all}$$
$$= (\pi/16)(1.5)^3(8,000)$$
$$= 5,301\,\text{in.-lb}$$

$$n = \text{HP} \times 63,025/T_{max}$$
$$= 15 \times 63,025/5,301 = 178\,\text{rpm}$$

Example 7.13 Design a solid steel shaft to transmit 20 hp at a speed of 1,500 rpm if the allowable shearing stress is 5 ksi.

Solution

$$\text{HP} = Tn/63,025$$

$$T = 20 \times 63,025/1,500 = 840\,\text{in.-lb.}$$

$$T_{max} = (\pi/16)d^3\tau_{all}$$

$$840 = (\pi/16)d^3(5,000)$$

$$d = 0.95\,\text{in.}$$

A 1-in. shaft should be used.

Example 7.14 Determine the maximum shearing stress in a 2-in. diameter solid shaft used to transmit 85 hp at a speed of 1,000 rpm.

$$\text{HP} = Tn/63,025$$

$$85 = T(1,000)/63,025$$

$$T = 5,357\,\text{in.-lb}$$

$$T = (\pi/16)d^3\tau_{all}$$

$$5,357 = (\pi/16)(2)^3\tau_{all}$$

$$\tau_{all} = 3,410\,\text{psi}$$

Metric System Problems

As discussed earlier, the unit of power in the metric system is

$$1\,\text{W} = 1\,\text{N.m/s} = 1\,\text{J/s}$$

Power was also defined as work per unit time

$$P = W/\text{min} = 2\pi Tn$$

This equation is valid for both measurement systems, English and metric. However, there are some changes in terms of metric units that must be done in order to make the metric system work. The unit of torque must be N.m, and rpm (n) will be divided by 60 to become rps. The equation for the metric system, then, is

$$P = 2\pi T n/60$$

where P is in watts, T is in N.m, and n is rpm

Example 7.15 Determine the maximum power that can be transmitted by a 20-mm solid shaft that is turning at 20 r/s if the allowable shear stress on the shaft is 75 MPa.

Solution

$$
\begin{aligned}
T_{max} &= (\pi/16)d^3\,\tau_{all}\\
&= (\pi/16)(0.02)^3\,(75\times10^6)\\
&= 117.81\,\text{N.m}
\end{aligned}
$$

$$
\begin{aligned}
P &= 2\pi T n/60\\
&= 2\pi(117.81)(20\times60)/60\\
&= 14.8\,\text{kW}
\end{aligned}
$$

Example 7.16 Determine the maximum shearing stress in a 30-mm diameter shaft that transmits 120 kW of power and is turning at 15 r/s.

$$
\begin{aligned}
P &= 2\pi T n/60\\
T &= P\times60/2\pi n\\
&= (120,000)60/(2\pi\times15\times60)\\
&= 1,273\,\text{N.m}
\end{aligned}
$$

$$
\begin{aligned}
T_{max} &= (\pi/16)d^3\,\tau_{all}\\
\tau_{all} &= (1,273)\times16/\pi(0.03)^3\\
&= 240\,\text{MPa}
\end{aligned}
$$

Chapter Summary

When the torque, or twisting moment, is applied to a circular shaft, it tends to rotate the shaft. This action causes torsional shear stresses that develop in the shaft. In a solid circular shaft, these stresses vary linearly from the center of the shaft to a maximum at the outer surface of the shaft. This is expressed as

$$\tau_{max} = 16T/\pi d^3$$

where τ_{max} is the torsional shearing stress in psi or MPa; T is the torque in in-lb or N.m, and d is the diameter of the shaft in inches or meters. The maximum resisting torque can be calculated from

$$T_{max} = (\pi/16)d^3 \tau_{all}$$

where τ_{all} is the allowable shearing stress in the shaft. The solid circular shaft can be designed from

$$d = (16T/\pi \tau_{all})^{1/3}$$

The maximum torsional shearing stress in a hollow circular shaft can be calculated from

$$\tau_{max} = 16T d_o/\pi(d_o^4 - d_i^4)$$

where d_o is the outer diameter and d_i is the inner diameter.

Rotating shafts can be used for transmitting power. The unit of transmitting power in the English system is horsepower (hp) and can be calculated from

$$HP = Tn/63,025$$

where T is the applied torque and n is rpm.

Review Questions

1. What is *torque*?
2. What is the definition of twisting moment?
3. Show the formula that can be used to calculate the twisting moment on a solid circular shaft.
4. What is the definition of torsional shearing stress in a solid circular shaft?
5. Show the formula that can be used to calculate the torsional shearing stress given the applied torque on the shaft.
6. How would you calculate the torsional shearing stress in a hollow circular shaft?
7. How would you calculate the horsepower developed by a rotating shaft?
8. How would you calculate the applied torque on a rotating shaft given the transmitting power?
9. Show the formula that can be used to calculate the transmitting power in the metric system of units.
10. What is the difference between rpm and rps, and where can they be applied?

Problems

1. Determine the developed torque on an automobile engine with 110 hp power at 5,000 rpm.
2. Design a solid circular shaft operating at 2,000 rpm to transmit 10 hp if the allowable shearing stress is 6,000 psi.
3. Calculate the torque that must be applied to a solid circular shaft 100 mm in diameter if the maximum torsional shearing stress is 60 MPa.
4. Calculate the maximum torsional shearing stress in a hollow circular shaft 4 in. outer diameter and 3 in. inner diameter if the developed torque on the shaft is 20,000 in. lb.
5. Calculate the maximum torsional shearing stress in a hollow circular shaft 15 mm outer diameter and 10 mm inner diameter if the developed torque on the shaft is 15,000 N.mm.
6. Calculate the maximum torsional shearing stress in a 2 in. diameter solid circular shaft that transmit 30 hp at 500 rpm.
7. A 1½ in. diameter solid circular shaft transmits 15 hp with an allowable shearing stress of 10,000 psi. Find the rpm.
8. A 50 mm diameter solid circular shaft transmits the power at 400 rpm. Find the maximum horsepower if the allowable shearing stress is 50 MPa.
9. Calculate the horsepower that a solid circular shaft 5 in. in diameter transmits at 200 rpm if the allowable shearing stress is 7,000 psi.
10. Calculate the developed torque on a hollow solid circular shaft 3 in. outer diameter and 2 in. inner diameter if the allowable shearing stress is 8,000 psi.

Shear and Bending Moment in Beams

8

Overview

A beam is one of the most common structural pieces loaded by forces acting perpendicular to its longitudinal axis. It is supported at its ends, where it supports the loads. The study of beams is important to a designer; designers must carefully study the behavior of the beam and choose the material and its dimensions so that the beam will safely carry the loading. In this chapter we will show how to evaluate shear force and bending moment in the beam and how to use this determination to design a beam. Shear force and bending moment are developed internally by external loading and reactions.

Learning Objectives

Upon completion of this chapter, you will be able to define beam and loading and calculate shear force and bending moment in the beam at different cross sections. You will also learn how to construct shear and moment diagrams for various beam loadings. Your knowledge, application, and problem solving skills will be determined by your performance on the chapter test.

8.1 Types of Beams

A beam is a member which is long compared to its cross-sectional dimension, and is loaded by the forces perpendicular to its long dimension. The beam is in equilibrium under the action of an applied system of forces and the reactions. Beams can be classified into types by the number and position of supports.

1. *Simply supported beam*: If the beam supports at the ends are either pins or rollers, the beam is called *simply supported* or a *simple beam*. Note that there is no restraint against rotation or translation at the supports (Fig. 8.1).

© Springer International Publishing Switzerland 2015
P. Ghavami, *Mechanics of Materials*, DOI 10.1007/978-3-319-07572-3_8

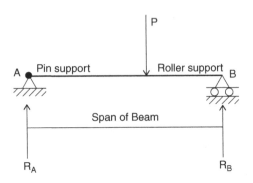

Fig. 8.1

2. *Fixed beam*: A beam is called a *fixed beam* if it is supported by two fixed supports that do not allow any end rotation or translation as the beam is loaded (Fig. 8.2).

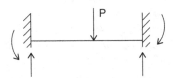

Fig. 8.2

3. *Cantilever beam*: A beam with a fixed support at one end and the other end unsupported is called a cantilever beam. Under loading, the fixed support does not allow rotation or translation (Fig. 8.3).

Fig. 8.3

4. *Overhanging beam:* A beam with one or two supports that are not positioned at the ends of the beam is called an overhanging beam (Fig. 8.4)

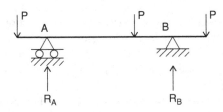

Fig. 8.4

5. *Continuous beam*: A beam that rests continuously over three or more supports is
   called a continuous beam (Fig. 8.5).

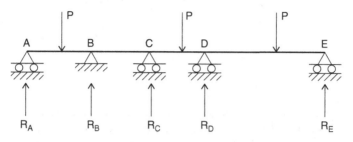

Fig. 8.5

The simply supported, cantilever, and overhanging beams are grouped as *stati-cally determinant*, because their reactions can be calculated using three laws of equilibrium:

$\sum F_x = 0$, $\sum F_y = 0$, and $\sum M = 0$. The fixed and continuous beams are grouped as *statically indeterminate*, because their reactions cannot be calculated only by using the above equations.

In this case, an additional equation based on the deformation situation of the beam must be established.

8.2   Types of Loadings

Loads on beams are classified as concentrated or distributed.

1. *Concentrated load*: This load acts over a short length of a beam and at a definite point. The unit of concentrated load in the English system is pounds or kips, and in the metric system is newtons (Fig. 8.6).

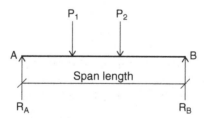

Fig. 8.6

2. *Distributed load:* This load spreads out over a considerable length of a beam. One form of distributed load is called a uniformly distributed load, which exists over a partial or the entire length of a beam (Fig. 8.7)

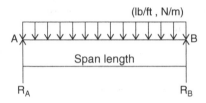

Fig. 8.7

In this type of loading, the load density is the same for the entire length of loading. For example, if $L = 18$ ft, and w (density) $= 300$ lb-ft, then the total load is 300 lb-ft $\times$ 18 ft $= 5,400$ lb. The unit for load density in the English system is pounds per linear foot (lb/ft, or kips per linear foot (kips/ft), and in the metric system is newtons per meter (N/m). If the load is non-uniformly distributed, then the beam has a varying density. In this case, the distributed load diagram looks like a triangle (Fig. 8.8) or a trapezoid shape.

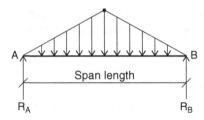

Fig. 8.8

In engineering design and applications, the loads may be classified as *dead loads* or *live loads*. Dead loads are permanent static loads which are imposed on a structure, such as the weight of the structural members, materials, and equipment. Live loads are movable loads such as vehicles and people or natural forces like ice, wind, earthquake, snow, or any other impacts that influence motion on a structure.

8.3   Beam Reactions

To analyze and compute the internal forces along the beam, first it is necessary to find the beam reactions. To determine the beam reactions we must apply the three laws of equilibrium as we studied earlier. These laws stated:

> The algebraic sum of all forces in the horizontal and vertical directions must be equal to zero. Also, the algebraic sum of moments in respect to any point must be equal to zero.

This, mathematically will be written as:

$$\sum F_h = 0 \tag{8.1}$$

$$\sum F_v = 0 \tag{8.2}$$

$$\sum M = 0 \tag{8.3}$$

The first and second equations above assure that the beam will not move in the horizontal or vertical direction. The third law assures that the beam will not rotate with respect to any point in the plane of action.

Example 8.1 Find the reactions for the simply supported beam AB shown in Fig. 8.9.

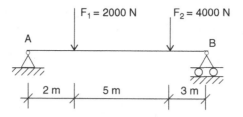

Fig. 8.9

Solution We write the first law of equilibrium

$$\sum F_h = 0$$

$$R_h = 0 \quad \text{at the pin support}$$

(Note: there is no horizontal force at the roller support)
Now we write the second law of equilibrium

$$\sum F_v = 0$$

$$R_A - F_1 - F_2 + R_B = 0$$

$$R_A - 2,000 - 4,000 + R_B = 0$$

$$R_A + R_B = 6,000\,\text{N}$$

The last equation has two unknowns and cannot be solved for reactions R_A or R_B. Therefore, we need another equation, which is the equation of equilibrium.
The moment equation with respect to point B is

$$R_A(10) - 2,000(8) - 4,000(3) + R_B(0) = 0$$

$$10R_A - 16,000 - 12,000 = 0$$

$$10R_A = 28,000$$

$$R_A = 2,800\,\text{N}$$

Knowing $R_A = 2,800$ N, we can calculate R_B from

$$R_A + R_B = 6,000\,\text{N}$$

and

$$R_B = 3,200\,\text{N}$$

Example 8.2 A simply supported beam with a span of 10 ft is under a uniformly distributed load of 200 lb/ft (Fig. 8.10). Compute the reactions at the supports A and B.

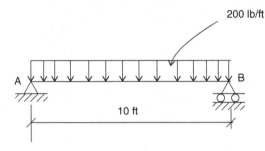

Fig. 8.10

Solution The total distributed load on the beam is

$$F = 200\,\text{lb/ft} \times 10\,\text{ft} = 2,000\,\text{lb}$$

We can replace the total uniformly distributed load by an equivalent concentrated load of 2,000 lb acting at the midspan of the beam. Now we can consider the beam the same as in the last example for a concentrated load of 2,000 lb (Fig. 8.11).

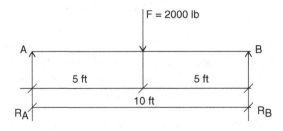

Fig. 8.11

We write the first law of equilibrium

$$\sum F_h = 0$$

$$R_h = 0 \quad \text{at the pin support}$$

(Note: there is no horizontal force at the roller support)
Now we write the second law of equilibrium

$$\sum F_v = 0$$

$$R_A - F + R_B = 0$$

$$R_A - 2,000 + R_B = 0$$

$$R_A + R_B = 2,000\,\text{lb}$$

The last equation has two unknowns and cannot be solved for reactions R_A or R_B. Therefore, we need another equation, which is the equation of equilibrium.
The moment equation with respect to point B is

$$R_A(10) - 2,000(5) + R_B(0) = 0$$

$$10R_A = 10,000$$
$$R_A = 1,000\,\text{lb}$$

Knowing $R_A = 1,000$ lb, we can calculate R_B from

$$R_A + R_B = 2,000\,\text{lb}$$

and

$$R_B = 1,000\,\text{lb}$$

Example 8.3 Find the reactions for the simply supported beam AB shown (Fig. 8.12a).

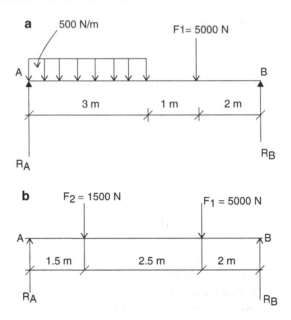

Fig. 8.12

Solution Now we are dealing with one concentrated load of $F_1 = 5,000$ N and a partial, uniformly distributed load of 500 N/m. First, we calculate the total uniformly distributed load and then replace it by an equivalent concentrated load.

$$F_2 = 500\,\text{N/m} \times 3\,\text{m} = 1,500\,\text{N}$$

The beam now has two concentrated loads as shown (Fig. 8.12b) and the reactions will be calculated the same way we did in Example 8.1.

We write the first law of equilibrium

$$\sum F_h = 0$$

$$R_h = 0 \quad \text{at the pin support}$$

(Note: there is no horizontal force at the roller support)
Now we write the second law of equilibrium

$$\sum F_v = 0$$

$$R_A - F_1 - F_2 + R_B = 0$$

$$R_A - 5,000 - 1,500 + R_B = 0$$

$$R_A + R_B = 6,500\,\text{N}$$

The last equation has two unknowns and cannot be solved for reactions R_A or R_B. Therefore, we need another equation, which is the equation of equilibrium.

The moment equation with respect to point B is

$$R_A(6) - 1,500(4.5) - 5,000(2) + R_B(0) = 0$$

$$6R_A - 6,750 - 10,000 = 0$$

$$6R_A = 16,750$$

$$R_A = 2,792\,\text{N}$$

Knowing $R_A = 2,792$ N, we can calculate R_B from

$$R_A + R_B = 6,500\,\text{N}$$

and

$$R_B = 3,708\,\text{N}$$

Example 8.4 A simply supported beam with a span of 20 ft is under a uniformly distributed load of 1,200 lb/ft and a concentrated load of 5,000 lb (Fig. 8.13a). Compute the reactions at the supports A and B.

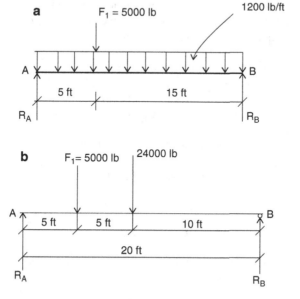

Fig. 8.13

Solution Now we are dealing with one concentrated load of $F_1 = 5,000$ lb, and a uniformly distributed load of 1,200 lb/ft. First, we calculate the total uniformly distributed load and then replace it with an equivalent concentrated load.

$$F_2 = 1,200 \, \text{lb/ft} \times 20 \, \text{ft} = 24,000 \, \text{lb}$$

The beam now has two concentrated loads shown (Fig. 8.13b) and the reactions will be calculated the same way we did it in Example 8.1.

We write the first law of equilibrium

$$\sum F_h = 0$$

$$R_h = 0 \quad \text{at the pin support}$$

(Note: there is no horizontal force at the roller support)
Now we write the second law of equilibrium

$$\sum F_v = 0$$

$$R_A - F_1 - F_2 + R_B = 0$$

$$R_A - 5,000 - 24,000 + R_B = 0$$

$$R_A + R_B = 29,000 \, \text{lb}$$

The last equation has two unknowns and cannot be solved for reactions R_A or R_B. Therefore, we need another equation, which is the equation of equilibrium.

The moment equation with respect to point B is

$$R_A(20) - 5,000(15) - 24,000(10) + R_B(0) = 0$$
$$20R_A - 75,000 - 240,000 = 0$$
$$20R_A = 315,000$$
$$R_A = 15,750 \, \text{lb}$$

Knowing $R_A = 15,750$ lb, we can calculate R_B from

$$R_A + R_B = 29,000 \, \text{lb}$$

and

$$R_B = 13,250 \, \text{lb}$$

Example 8.5 Find the reactions for the overhanging beam shown (Fig. 8.14a).

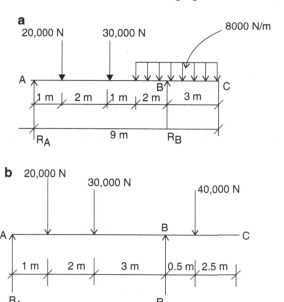

Fig. 8.14

Solution Now we are dealing with two concentrated loads $F_1 = 20,000$ N and $F_2 = 30,000$ N, and a partial, uniformly distributed load of 8,000 N/m. First, we calculate the total uniformly distributed load and then replace it with an equivalent concentrated load.

$$F_3 = 8,000\,\text{N/m} \times 5\,\text{m} = 40,000\,\text{N}$$

The beam now has three concentrated loads as shown (Fig 8.14b) and the reactions will be calculated the same way we did it in Example 8.1.

We write the first law of equilibrium

$$\sum F_\text{h} = 0$$

$$R_\text{h} = 0 \quad \text{at the pin support}$$

(Note: there is no horizontal force at the roller support)
Now we write the second law of equilibrium

$$\sum F_\text{v} = 0$$

$$R_A - F_1 - F_2 - F_3 + R_B = 0$$

$$R_A - 20,000 - 30,000 - 40,000 + R_B = 0$$

$$R_A + R_B = 90,000\,\text{N}$$

The last equation has two unknowns, and cannot be solved for reactions R_A or R_B. Therefore, we need another equation, which is the equation of equilibrium. The moment equation with respect to point B is

$$R_A(6) - 20,000(5) - 30,000(3) + 40,000(0.5) + R_B(0) = 0$$
$$6R_A - 170,000 = 0$$
$$R_A = 28,333\,\text{N}$$

Knowing $R_A = 28,333$ N, we can calculate R_B from

$$R_A + R_B = 90,000\,\text{N}$$

and

$$R_B = 61,667\,\text{N}$$

Practice Problems
For Problems 1 through 10, calculate the reactions at points A and B for beams loaded as shown in Figs. 8.15–8.24.

1.

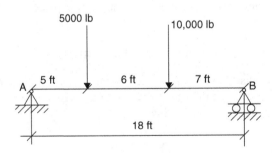

Fig. 8.15

2.

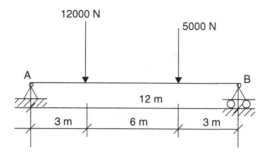

Fig. 8.16

3.

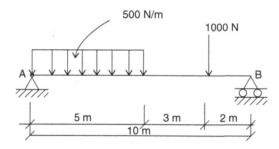

Fig. 8.17

4.

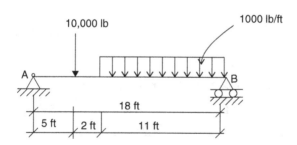

Fig. 8.18

5.

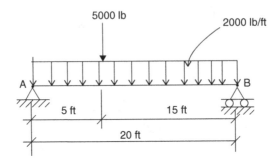

Fig. 8.19

6.

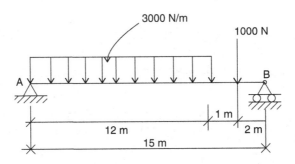

Fig. 8.20

7.

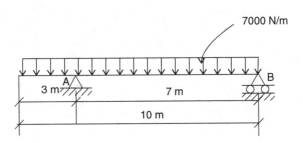

Fig. 8.21

8.

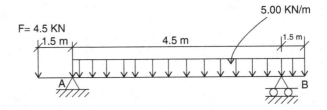

Fig. 8.22

9.

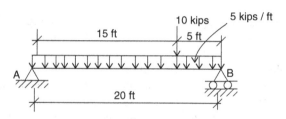

Fig. 8.23

10.

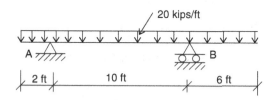

Fig. 8.24

8.4   Shear Force

When a beam is subjected to applied forces, internal forces develop in the beam. Since the beam has supports and the system is in an equilibrium condition, the beam should resist these internal forces. This type of force is called *shear force*, or simply shear. The magnitude of the shear has a direct effect on the magnitude of the shear stress in the beam, and also on the design and analysis of the structural members.

Shear forces develop along the entire length of the beam, and their magnitude varies at each cross section of the beam. They actually act perpendicularly to the longitudinal axis of the beam. Our main objectives in finding the shear in beams are twofold. First, we have to know the maximum value of the shear in order to design the beam properly so that it can resist overall shear and will not fail when it is loaded.

The second objective is to calculate the values of shear along the length of the beam, so that we can find out where the beam fails under bending and where the shear value is zero. This will be clearly explained later in the section of shear/bending moment diagrams.

8.5   Computation of Shear Force

As mentioned earlier, the designer must first know the magnitude of the shear force at any section of the beam to get the right picture of beam design. The magnitude of shear force at any section of the beam is equal to the sum of all the vertical forces either to the right or to the left of the section. To compute the shear force for any section of the beam, we simply call the upward forces (reactions) positive and downward forces (loads) negative. Then, *the magnitude of the shear force at any section of the beam is equal to sum of the reactions minus the sum of the loads to the left of the section.* This can be stated as:

$$\text{Shear} = \text{Reactions} - \text{Loads}$$

To understand this concept, let's try an example.

Example 8.6 Find the shear force at sections C, D, and E for the simply supported beam shown (Fig. 8.25).

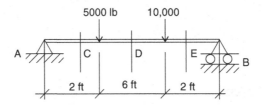

Fig. 8.25

Solution First, we calculate the reactions.

$$R_A + R_B = 15,000\,\text{lb}$$

Using the moment equilibrium equation

$$R_A(10) - 5,000(8) - 10,000(2) = 0$$
$$10R_A = 60,000$$
$$R_A = 6,000\,\text{lb}$$

and

$$R_B = 9,000\,\text{lb}$$

Point C We usually represent a shear force by V, and we designate this shear force at C as V_c.

$$V_c = +6,000 - 0 = +6,000\,\text{lb}$$

Point D

$$V_d = +6,000 - 5,000 = +1,000 \, \text{lb}$$

Point E

$$V_e = +6,000 - 5,000 - 10,000 = -9,000 \, \text{lb}$$

Example 8.7 Compute the shear force (a) at 5 m and (b) at 15 m from the left end of the beam shown (Fig. 8.26).

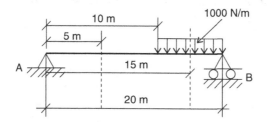

Fig. 8.26

Solution First we calculate the reactions.

The beam is loaded with a uniform distributed load of 1,000 N/m. The total distributed load is

$$W = 1,000 \, \text{N/m} \times 10 \, \text{m} = 10,000 \, \text{N}$$

$$R_A + R_B = 10,000 \, \text{N}$$

Using the moment equilibrium equation

$$R_A(20) - 10,000(5) = 0$$
$$20 R_A = 50,000$$
$$R_A = 2,500 \, \text{N}$$

and

$$R_B = 2,500 \, \text{N}$$

then

$$V_{5\,m} = +2,500 \, \text{N}$$
$$V_{15\,m} = +2,500 - 1,000 \, \text{N/m}(5 \, \text{m}) = 2,500 - 5,000 \, \text{N} = -2,500 \, \text{N}$$

Practice Problems

1. Compute the shear force at (a) 4 ft, (b) 12 ft, and (c) 16 ft from the left end of the beam shown (Fig. 8.27).

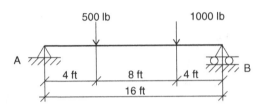

Fig. 8.27

2. Compute the shear force at (a) 4 ft, (b) 12 ft, and (c) 16 ft from the left end of the beam shown (Fig. 8.28).

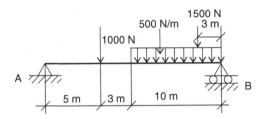

Fig. 8.28

3. Compute the shear force at (a) 2.5 m, (b) 5.5 m, and (c) 6.5 m from the left end of the beam shown (Fig. 8.29).

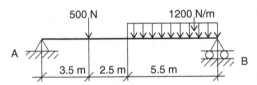

Fig. 8.29

4. Compute the shear force at (a) 5 ft, (b) 10 ft, and (c) 12 ft from the left end of the beam shown (Fig. 8.30).

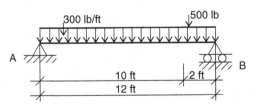

Fig. 8.30

5. Compute the shear force at (a) 1.5 m, (b) 2.5 m, and (c) 3.5 m from the left end of the beam shown (Fig. 8.31).

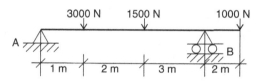

Fig. 8.31

6. Compute the shear force (a) at 2 m, (b) at 4 m and (c) at 8 m from the left end of the beam shown (Fig. 8.32).

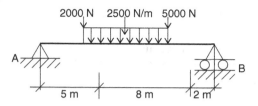

Fig. 8.32

8.6   Shear Diagrams

In the previous section, we computed the value of shear at any section along the length of a beam. Designing a beam requires the variation of shear at arbitrary locations on the beam. This important task can be accomplished using a graphical technique referred to as a *shear diagram*.

Shear diagrams are graphical representations of the shear force variations over the length of a beam. The shear diagram shows the magnitude of shear at various points along a beam and is useful in finding the maximum value of shear when designing a beam.

The shear diagram is drawn directly underneath the free body diagram of the beam and will show the variation of shear values over the length of the beam. The following examples illustrate the construction of shear diagrams for a variety of beam loadings.

Example 8.8 Construct a shear diagram for the simply supported beam shown (Fig. 8.33).

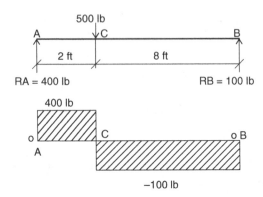

Fig. 8.33

Solution First we calculate the reactions. Using the laws of equilibrium, we find that $R_A = 400$ lb and $R_B = 100$ lb. Remember, to calculate the shear at any section, we use the equation

$$V = R - L$$

where V is the shear, R is the reaction, and L is the load at the left side of the section. Now we proceed to calculate the shear from reaction A until we reach a load of 500 lb. We will see that the shear value does not change and follows a horizontal straight line with a constant value of

$$V = +400 - 0 = +400\,\text{lb}$$

At the load 500 lb, the shear suddenly passes from positive to negative quantity, because

$$V = +400 - 500 = -100\,\text{lb}$$

Obviously, the shear goes to zero on the baseline when it suddenly passes from a positive to a negative quantity. From this point on until it reaches reaction B, the shear again stays constant on the horizontal line and its value is $V = -100$ lb. Notice that the value of the shear is represented by the distance from the baseline to the shear diagram (Fig. 8.34).

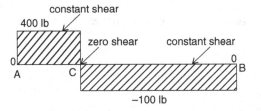

Fig. 8.34

Example 8.9 Construct a shear diagram for the simply supported beam shown (Fig. 8.35).

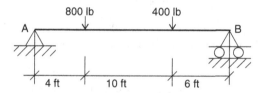

Fig. 8.35

Solution First, we calculate the reactions. Using the laws of equilibrium, we find that $R_A = 760$ lb and $R_B = 440$ lb. Working from the left end, as in the previous example, the shear values at the following sections are

$$
\begin{aligned}
V_{x=1\,\text{ft}} &= 760 - 0 = 760\,\text{lb} \\
V_{x=5\,\text{ft}} &= 760 - 800 = -40\,\text{lb} \\
V_{x=15\,\text{ft}} &= 760 - 800 - 400 = -440\,\text{lb}
\end{aligned}
$$

Plotting the positive shear values above the baseline and the negative shear values below the baseline results in the graphical representation shown in Fig. 8.36.

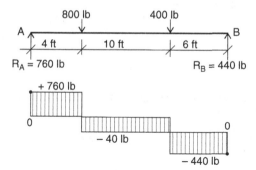

Fig. 8.36

Example 8.10 A 14-ft beam is loaded with a uniformly distributed load of 400 lb/ft as shown (Fig. 8.37). Construct a shear diagram for this beam.

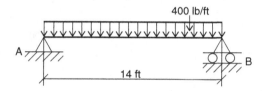

Fig. 8.37

Solution The total load is

$$400\,\mathrm{lb/ft} \times 14\,\mathrm{ft} = 5,600\,\mathrm{lb}$$

Calculation of reactions gives: $R_A = 2,800$ lb and $R_B = 2,800$ lb. Working from the left end as in the previous example, the shear values at the random sections are

$$
\begin{aligned}
V_{x=1\,\mathrm{ft}} &= +2,800 - (1 \times 400) = +2,400\,\mathrm{lb} \\
V_{x=4\,\mathrm{ft}} &= +2,800 - (4 \times 400) = +1,200\,\mathrm{lb} \\
V_{x=7\,\mathrm{ft}} &= +2,800 - (7 \times 400) = 0 \\
V_{x=14\,\mathrm{ft}} &= +2,800 - (14 \times 400) = -2,800\,\mathrm{lb}
\end{aligned}
$$

Notice that the shear decreases uniformly as x increases. At the center of the span ($x = 7$ ft) the shear goes to zero. Plotting the positive shear values above the baseline and the negative shear values below the baseline results in a shear diagram that resembles a sloping straight line (Fig. 8.38).

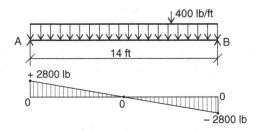

Fig. 8.38

Example 8.11 Construct the shear diagram for the cantilever beam loaded as shown in Fig. 8.39.

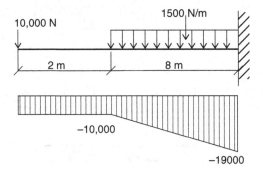

Fig. 8.39

Solution Notice that there is no reaction to the left of the beam. Working from the left end as in the previous example, the shear values at the random sections are

$$
\begin{aligned}
V_{x=2\,m} &= -10,000\,N \\
V_{x=3\,m} &= -10,000\,N - (1,500\,N/m \times 1\,m) = -11,500\,N \\
V_{x=10\,m} &= -10,000\,N - (1,500\,N/m \times 6\,m) = -19,000\,N \\
V_{max} &= -19,000\,N
\end{aligned}
$$

Plotting the positive shear values above the baseline and the negative shear values below the baseline results in the graphical representation of the shear values.

Practice Problems

1. For the beams and loadings shown in the following figures, construct the shear diagrams and find the magnitudes of maximum and minimum shear at the indicated sections.

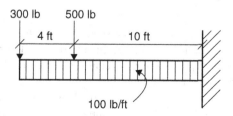

Fig. 8.40

2.

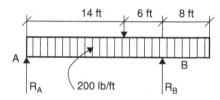

Fig. 8.41

3.

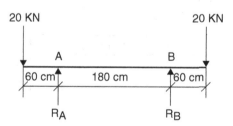

Fig. 8.42

4.

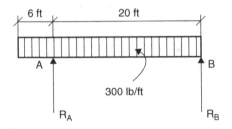

Fig. 8.43

5.

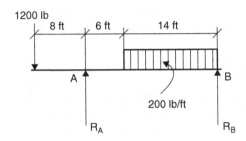

Fig. 8.44

6.

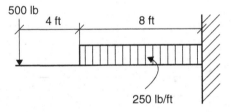

Fig. 8.45

8.7   Bending Moment

In the last two sections, we discussed the shear force and shear diagram and their important impact on designing a beam. In addition to these, there is another important internal quantity that must be considered before the design stage can proceed. This internal quantity is called the *bending moment*.

The bending moment at any section along the length of a beam is the measure of the tendency of the beam to bend due to the forces acting on it. *The bending moment will vary, but at any section of a beam is equal to the algebraic sum of the moments of the forces to either the right or to the left of the section.*

8.8   Computation of Bending Moment

The bending moment at any section along the length of a beam equals the moments of the reactions minus the moments of the loads to the left of the section under consideration.

$$M = M_R - M_L$$

Example 8.12 Determine the magnitude of the bending moment at the section 5 m from the left reaction of the beam shown in Fig. 8.46.

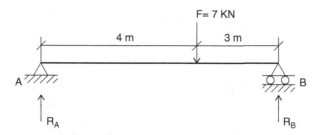

Fig. 8.46

Solution Using the equilibrium law, we can calculate the reaction $R_A = 3$ KN, and $R_B = 4$ KN.

$$
\begin{aligned}
M_{x=5\,m} &= M_R - M_L \\
&= R_A(5) - F(1) \\
&= 3(5) - 7(1) = +8\,\text{KN.m}
\end{aligned}
$$

Example 8.13 Determine the magnitude of the bending moment at the sections 0, 4, 6, and 14 ft from the left reaction of the beam shown in Fig. 8.47.

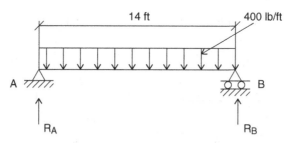

Fig. 8.47

Solution Using the equilibrium laws, we can calculate the reactions $R_A = R_B = 2,800$ lb.

The total load is

$$W = 400\,\text{lb/ft} \times 14\,\text{ft} = 5,600\,\text{lb}$$

$$
\begin{aligned}
M_{x=0\,\text{m}} &= 0 \\
M_{x=4\,\text{m}} &= 2,800(4) - (400 \times 4 \times 2) = +8,000\,\text{lb-ft} \\
M_{x=6\,\text{m}} &= 2,800(6) - (400 \times 6 \times 3) = +9.600\,\text{lb-ft} \\
M_{x=14\,\text{m}} &= 2,800(14) - (400 \times 14 \times 7) = 39,200 - 39,200 = 0
\end{aligned}
$$

Example 8.14 For the beam shown in Fig. 8.48, determine the magnitude of the bending moment at sections C and D.

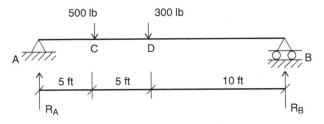

Fig. 8.48

Solution Using the equilibrium laws, we can calculate the reactions $R_A = 525$ lb and $R_B = 275$ lb.

$$
\begin{aligned}
M_C &= 525 \times 5 = 2,625\,\text{lb-ft} \\
M_D &= 525 \times 10 - 500 \times 5 = 2,750\,\text{lb-ft}
\end{aligned}
$$

Practice Problems

1. For the beam shown in Fig. 8.49, determine the bending moment at sections B and C.

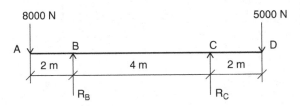

Fig. 8.49

2. For the beam shown in Fig. 8.50, determine the bending moment at sections C, D, and E.

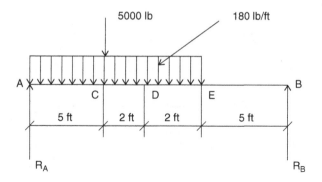

Fig. 8.50

3. For the beam shown in Fig. 8.51, determine the bending moment at sections C and D.

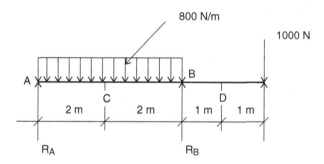

Fig. 8.51

4. For the uniformly distributed beam shown in Fig. 8.52, determine the bending moment at sections 6, 10, and 20 ft from the left section of the beam.

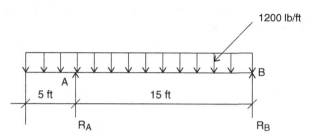

Fig. 8.52

5. For the simply supported beam with three concentrated loads shown in Fig. 8.53, determine the bending moment at sections 5, 9, and 18 ft from the left section of the beam.

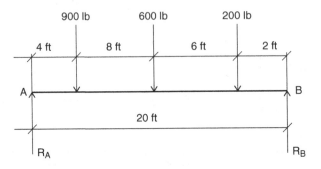

Fig. 8.53

6. For the simply supported beam with a uniformly distributed load shown in Fig. 8.54, determine the bending moment at the sections 10, 15, and 20 m from the left section of the beam.

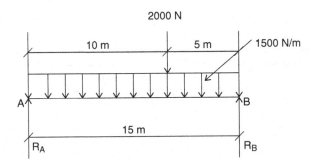

Fig. 8.54

7. For the beam shown in Fig. 8.55, determine the bending moment at the sections 60, 180, and 250 cm from the left section of the beam. Make sure to do the conversion into the metric system.

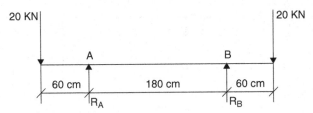

Fig. 8.55

8. For the beam with a partial, uniformly distributed load shown in Fig. 8.56, determine the bending moment for the sections at 6 f. and 15 ft from the end.

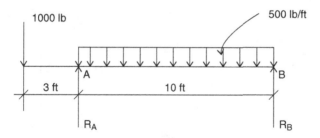

Fig. 8.56

9. For the beam with a partial, uniformly distributed load shown in Fig. 8.57, determine the bending moment for the sections 3 m, 7 m, and 8 m from the end.

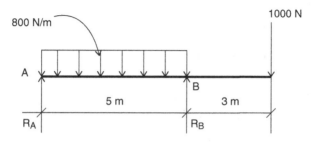

Fig. 8.57

8.9   Bending Moment Diagrams

A bending moment diagram is a diagram that shows the variations of bending moment values at each cross section along a beam. A bending moment diagram like the shear diagram discussed earlier is an important item in designing a beam. In fact, the designer must simultaneously consider the shear and bending moment diagrams, analyze them, and find out where the bending moment diagram reaches a maximum point.

8.10   Construction of a Bending Moment Diagram

To construct the bending moment diagram, we calculate the bending moment at the sections of interest along the beam, and then we draw the diagram below the shear diagram based on the calculated points of bending moments. The diagram does not have to be to scale; a good approximation will do the job. Notice that the positive bending moments are above the baseline, and the negative ones are below. Also, usually for the beams loaded with concentrated forces, the bending moment diagram is a straight line drawn between the forces.

Example 8.15 Draw the bending moment diagram for the beam shown in Fig. 8.58.

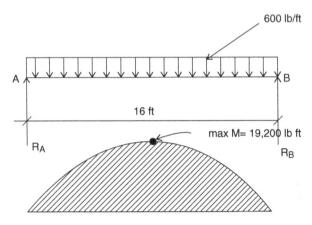

Fig. 8.58

Solution We calculate the reactions:

$$R_A + R_B = 9,600 \, \text{lb}$$

$$R_A(16) - 9,600(8) = 0, \quad R_A = 4,800 \, \text{lb}$$

$$R_A + R_B = 4,800 \, \text{lb}$$

Using the method explained previously we compute the values of the bending moment at several sections in order to construct the bending moment diagram

$$
\begin{aligned}
M &= M_R - M_L \\
M_{x=2\,\text{ft}} &= 4,800 \times 2 - 600 \times 2 \times 1 = 8,400 \, \text{lb-ft} \\
M_{x=4\,\text{ft}} &= 4,800 \times 4 - 600 \times 4 \times 2 = 14,400 \, \text{lb-ft} \\
M_{x=8\,\text{ft}} &= 4,800 \times 8 - 600 \times 8 \times 2 = 19,200 \, \text{lb-ft} \, (\text{maximum value!}) \\
M_{x=16\,\text{ft}} &= 4,800 \times 16 - 600 \times 16 \times 8 = 0
\end{aligned}
$$

We plot the above points and connect them to get the moment diagram. Notice that the value of maximum bending moment occurs at the section of the beam at which the shear is zero (this can be proven using the knowledge of calculus).

In this example, the maximum value of the bending moment is at a distance of 8 ft from the R_A. This value is 19,200 lb-ft. In designing the beams, the value of the maximum bending moment is the moment with which we particularly are concerned.

Example 8.16 Draw the bending moment diagram for the beam shown in Fig. 8.59 and find the maximum bending moment.

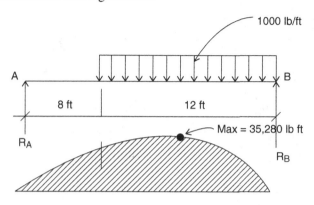

Fig. 8.59

Solution We calculate the reactions using the equilibrium laws

$$R_A + R_B \; 12,000 \text{ lb}$$

$$R_A(20) - 12,000(6) = 0$$

$$R_A = 3,600 \text{ lb}, \quad \text{and} \quad R_B = 8,400 \text{ lb}$$

Using the method explained previously

$$
\begin{aligned}
M &= M_R - M_L \\
M_{x=0\,\text{ft}} &= 0 \\
M_{x=8\,\text{ft}} &= 3,600 \times 8 = 28,800 \text{ lb-ft} \\
M_{x=12\,\text{ft}} &= 3,600 \times 12 - (4 \times 1,000 \times 2) = 35,200 \text{ lb-ft}
\end{aligned}
$$

The maximum bending moment occurs at the section where the shear passes through zero. To find this point, let us define x as the distance from reaction R_A. Then we have

$$3,600 - [(x - 8) \times 1,000] = 0$$

$$x = 11.6 \text{ ft}$$

and

$$M_{\max}(@x = 11.6\,\text{ft}) = 3,600 \times 11.6 - 3.6 \times 1,000(3.6/2)$$

$$M_{\max} = 35,280 \text{ lb-ft}$$

Example 8.17 Draw the shear and moment diagrams for the overhanging beam shown in Fig. 8.60, and find the maximum bending moment.

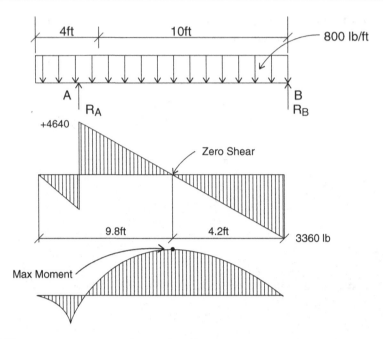

Fig. 8.60

Solution Using the laws of equilibrium, we find the reactions

$$R_A = 7,840\,\text{lb}$$

$$R_B = 3,360\,\text{lb}$$

Shear calculations at $x = 4$ ft, and $x = 14$ ft give

$$V_{x=4\,\text{ft}} = 7,840 - 3,200 = 4,640\,\text{lb}$$

$$V_{x=14\,\text{ft}} = 7,840 - 800 \times 14 = -3,360\,\text{lb}$$

As the shear diagram shows, the shear passes through zero at two points: $x = 4$ ft and

$$7,840 - 800x = 0, \quad x = 9.8\,\text{ft}$$

and

$$M_{\text{max}} = 7,840 \times 5.8 - 800 \times 9.8 \times 4.9 = 7,056\,\text{lb-ft}$$

The moment diagram shows the maximum and minimum value of the bending moment for above beam.

Practice Problems
Draw the moment diagram for the following beams and find the location of the maximum bending moment.

1.

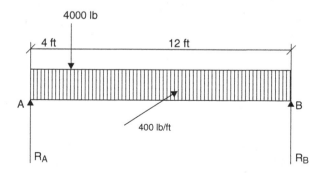

Fig. 8.61

2.

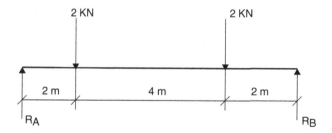

Fig. 8.62

3.

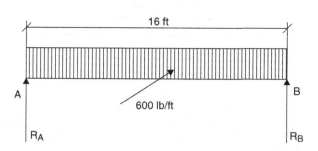

Fig. 8.63

4.

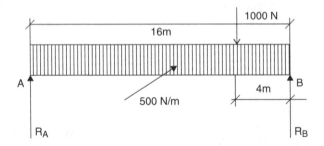

Fig. 8.64

5.

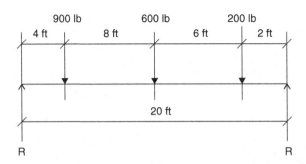

Fig. 8.65

Chapter Summary

Under different types of loading, internal forces develop in the beam. These forces are called shear forces. Shear force can be calculated at any arbitrary cross section of a beam using the sum of the reactions and loads at the left side of the beam. This was formulated as:

$$V = R - L$$

The graphical representation of the shear along the length of the beam is called a shear diagram. With the help of a shear diagram we can find the value of shear at any point along the beam. This method is very useful for helping a designer of beams understand the variations of the shear along the length of a beam.

The calculation of the bending moment and its importance in designing a beam can be shown with some simple examples. To calculate the bending moment, we simply take into consideration the moment of all forces (reactions and loads) with respect to any point to the left side of the point of interest. In fact, the bending moment at any section along the length of the beam is the measure of the tendency of the beam to bend due to the forces acting on it.

Now we can introduce the moment diagram, and we show how important that is for the designer to calculate both the values of shear and bending moment at any point along the length of the beam. It can also be proven mathematically or graphically that the maximum value of the bending moment will occur at the section where the shear is zero.

Review Questions

1. What is a beam?
2. What are the loads acting on a beam?
3. What are the reactions, and what laws can be applied to calculate them?
4. What is the shear in the beam?
5. How would you calculate the shear at any section of a beam?
6. What is a shear diagram? Explain how important it is in designing beams.
7. What is the bending moment? How would you calculate it at any section of a beam?
8. What is a bending moment diagram?
9. Where does the bending moment reach maximum value? Explain.
10. Explain how a shear diagram and a bending moment diagram together would help a designer come up with the best solution for beam design.

Problems

For the beams and loadings shown in the following figures, (a) find the reactions at the supports; (b) draw the complete shear diagram; and (c) draw the bending moment diagram and determine the magnitude and location of the maximum bending moment.

1.

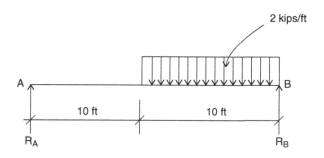

Fig. 8.66

2.

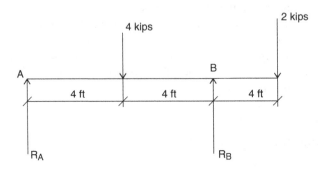

Fig. 8.67

3.

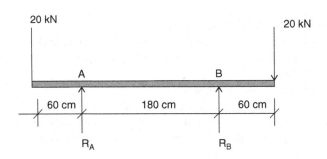

Fig. 8.68

4.

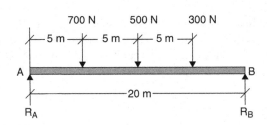

Fig. 8.69

5.

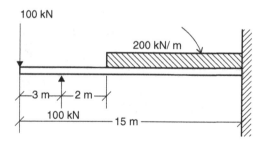

Fig. 8.70

6.

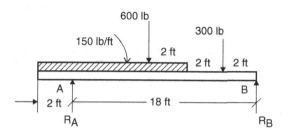

Fig. 8.71

7.

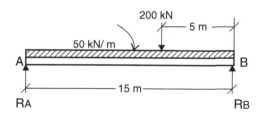

Fig. 8.72

8.

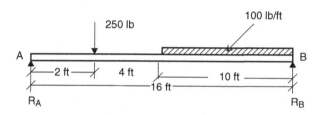

Fig. 8.73

9.

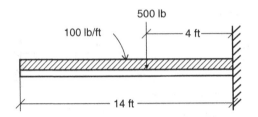

Fig. 8.74

10.

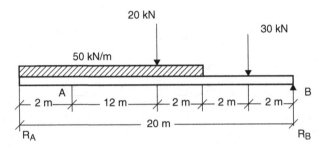

Fig. 8.75

Bending Stresses in Beams

9

Overview

In the previous chapter we discussed the shear forces and bending moments that were caused by applying external forces to a beam. Therefore, a beam must resist these shear forces and bending moments. The beam itself must develop internal resistance to (1) resist shear forces, referred to as shear stresses; and to (2) resist bending moments, referred to as bending stresses or flexural stresses. For beam design purposes, it is very important to calculate the shear stresses and bending stresses at the various locations of a beam. Since the shear stresses have been covered earlier, this chapter will only emphasize the bending stresses that are caused by bending moments.

Learning Objectives

Upon completion of this chapter, you will be able to define the neutral plane, called the plane of zero bending. You will also learn how to calculate the maximum bending stress using the flexure formula and compare it with the maximum resisting moment when the design problem arises. Your knowledge, application, and problem solving skill will be determined by your performance on the chapter test.

9.1   Introduction

When shear forces and bending moments develop in a beam because of external forces, the beam will create internal resistance to these forces, called *resisting shearing stresses* and *bending stresses*.

Consider the simply supported beam shown in Fig. 9.1 before it is subjected to the load. If we look at a short length of the beam between the cross sections A-B and C-D, we see that these two segments are parallel and of equal lengths. As the beam is loaded and bends, points A and C move closer together and points B and D are stretched at the bottom (Fig. 9.2).

© Springer International Publishing Switzerland 2015
P. Ghavami, *Mechanics of Materials*, DOI 10.1007/978-3-319-07572-3_9

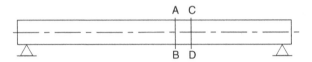

Fig. 9.1

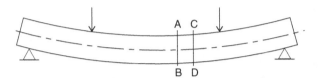

Fig. 9.2

In other words, segment AC shortens and segment BD lengthens. The top of the beam is in compression and the bottom is in tension. The maximum compression stress occurs at the upper fiber of the beam; the magnitude of the stress decreases below the upper fiber, and eventually goes to zero at the *neutral plane*. The intersection of the neutral plane with the cross-sectional plane is called the *neutral axis* (Fig. 9.3). Also, the maximum tensile stress occurs at the bottom fiber of the beam and goes to zero at the neutral plane.

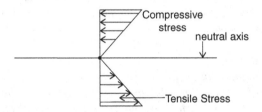

Fig. 9.3

The theory of beams plays an important role in structural design and is used by the designer as a powerful technique to analyze various structures. In fact, beam theory gives an accurate and deep understanding of the behavior of a structure. One of the most useful models of beam theory was developed by Euler and Bernoulli, known as the *Euler–Bernoulli theory of beams*. This theory was based on the following assumptions:

1. The cross section of the beam is infinitely rigid in its own plane.
2. The cross section of the beam remains in the plane after deformation.
3. The cross section of the beam remains normal to the deformed axis of the beam.

It has been assumed that the stress in any fiber does not exceed the proportional limit of the material, and so it will support Hooke's law—meaning that the stress is proportional to the distance from the neutral plane and the sections A-B and C-D, before and after bending, remain in a straight line (Fig. 9.4).

As a result, the stress distribution will have a triangular shape. Stress changes from zero at the neutral axis to either a maximum compressive stress at the top outer fiber or to a maximum tensile stress at the bottom outer fiber.

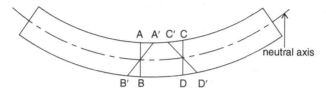

Fig. 9.4

9.2   Resisting Moment

The next question is how do tension and compression stress resist the externally applied bending moment. Consider a simple beam subjected to a concentrated load P (Fig. 9.5). The free-body diagram shows the part of the beam to the left of the cross section X-X. The reaction R_1 creates a clockwise rotation with respect to point O, which is the bending moment of the section. The next item is the moment of the stresses on the cross section.

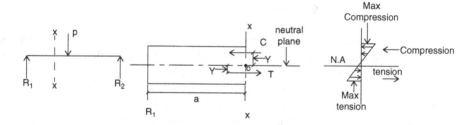

Fig. 9.5

If we call the sum of the compressive stresses C and the sum of the tensile stresses T, then the T-C *couple* is the *resisting moment* developed by the beam in response to the externally applied bending moment (Fig. 9.6).

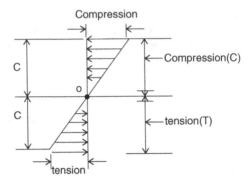

Fig. 9.6

This resisting moment tends to cause a counterclockwise rotation, and is equal to the bending moment if the beam is in equilibrium. Mathematically, this can be written as

$$R_1 \times a = C \times y + T \times y$$

This is called the *theory of flexure* in beams. Knowing the bending moment of the beam, we will be able to design the beam to resist bending and its resisting moment is equal to the external bending moment.

9.3   The Flexure Formula

As we mentioned earlier, if the stresses in the beam are lower than the material's proportional limit, Hooke's law can be applied; this means that the strains are related to the stresses by a constant called the modulus of elasticity. In other words, the bending stresses caused by bending moments are proportional to the distance from the neutral axis of the beam. The relationship between the bending moment (M) on the cross section, bending stresses (σ), and properties of the cross section, called the *flexure formula*, is applied for the purposes of analyzing and designing beams.

The maximum bending stress at the outer fiber with a distance c from the neutral axis is proportional to the bending stress at a distance y from the neutral axis. Figure 9.6 shows maximum tension or compression bending stress at the outer fiber. This can be written as

$$\sigma_y / y = \sigma_{max} / c \qquad\qquad (9.1)$$

$$\sigma_y = y \sigma_{max} / c$$

The internal resisting moment is the sum of the force F times lever arm y about the neutral axis

$$M = \sum F \cdot y \qquad (9.2)$$

where

$$F = a\,\sigma_y \quad (a \text{ is an infinitesimal area}) \qquad (9.3)$$

or

$$F = a\,\sigma_{max}(y/c)$$

and

$$M = \sum a\,\sigma_{max}(y/c) \cdot y = \sum a y^2 \sigma_{max}/c \qquad (9.4)$$

or

$$\sigma_{max} = Mc/\sum a y^2 = Mc/I \qquad (9.5)$$

where $I = \sum a\, y^2$ is the moment of inertia of a cross section about an x-x axis.

In Eq. 9.5 above, if the stress is in psi, distance is in inches, and the moment of inertia is in in.[4], then the unit of bending moment in the English system is lb-in. If we precisely look at Eq. 9.5 we will find out that it represents the resisting moment dealing with the size, material, and shape of the beam cross section. The expression can be simplified by substituting $S = I/c$, called the *section modulus*. Then the formula becomes

$$M = \sigma S \qquad (9.6)$$

The section modulus S is an important property of the cross section and can be considered a measure of the strength of a beam. It is also a characteristic of the bending capacity and stiffness of a beam cross section. The larger the section modulus, the smaller the stresses developed in the section for a certain bending moment. Equation 9.6 shows that the larger the section modulus, the larger the bending moment developed at the beam cross section for certain compressive or tensile stresses. Values of section moduli for some cross sectional shapes can be found in the tables of properties of materials.

Example 9.1 Calculate the maximum bending stress of the beam if the cross-sectional dimension is 5 in. × 10 in. and the bending moment is 50,000 lb-in. (Fig. 9.7).

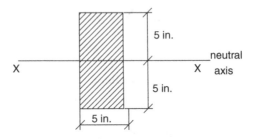

Fig. 9.7

Solution We make some assumptions to be able to solve the problem with the given information.

1. The centroid of a rectangle is at the center of the rectangle; this means that the neutral axis is at the center.
2. The moment of inertia for a rectangle shape with respect to the centroidal axis can be found from Chap. 5, and is $(1/12)\,bh^3$.

Using the flexure formula

$$\sigma = Mc/I$$

$$I = (1/12)\,bh^3 = (1/12)\left(5 \times 10^3\right) = 416.67\,\text{in.}^4$$

$$\sigma = (50,000 \times 5)/416.67$$

$$\sigma = 600\,\text{psi}$$

Example 9.2 A nominal size piece of structural timber 6 in. × 10 in. is subjected to a vertical loading (Fig. 9.8). Determine the section modulus of the beam if the maximum bending stress is established.

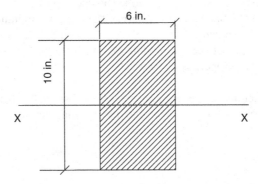

Fig. 9.8

Solution Assume that X-X is the axis of bending. Using the definition of the section modulus as

$$S = I/c$$
$$S = \left(bh^3/12\right)/h/2 = bh^3/6$$

Substituting the given data, we get

$$S = (5\tfrac{1}{2})(9\tfrac{1}{2})^2/6 = 82.7\,\text{in.}^3 \quad \text{(see table value of structural timbers!)}$$

Example 9.3 Calculate the maximum bending stress for a steel bar 45 mm in diameter (Fig. 9.9) if a bending moment of 200 N.m is applied.

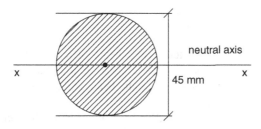

neutral axis

X   X

45 mm

Fig. 9.9

Solution Use the definition of section modulus as

$$\sigma = Mc/I$$

Calculate the moment of inertia of a circular area

$$I = \pi d^4/64 = \pi(0.045\,\text{m})^4/64 = 2 \times 10^{-7}\,\text{m}^4$$

$$\sigma = Mc/I = 200 \times 0.0225/\left(2 \times 10^{-7}\right) = 23\,\text{MPa}$$

Example 9.4 Calculate the width of a 10-in. deep rectangular beam with a bending moment of 120,000 lb-in. Assume that the allowable bending stress is 2,000 psi.

Solution Using the definition of the section modulus as

$$\sigma = Mc/I$$

or

$$\sigma = M/S$$

and

$$S = M/\sigma$$

$$S = 120,000\,\text{lb-in.}/2,000\,\text{psi}$$

$$S = 60\,\text{in.}^3$$

From Example 9.2, we found that the section modulus of a rectangular beam is

$$S = bh^2/6 = 60$$

Solving for b, it gives:

$$b = 60 \times 6/10^2 = 3.6\,\text{in.}$$

Example 9.5 A simply supported timber beam of length of 20 ft carries a uniformly distributed load of 300 lb/ft. Determine the maximum bending stress. Use Example 9.2 for the dimensional size of the beam and assume $S = 82.7$ in.^{3}.

Solution From Appendix A we find the maximum bending moment for a simple beam with a uniformly distributed load.

$$M = wl^2/8 = 300(20)^2/8 = 15,000\,\text{lb-ft}$$

The maximum stress is

$$\sigma = M/S = (15,000\,\text{lb-ft})(12)/82.7 = 2,176.4\,\text{psi}$$

Practice Problems

1. Calculate the maximum bending stress on a rectangular beam 12 in. × 20 in. that has an applied bending moment of 10,000 lb-in.
2. Calculate the maximum bending stress on a square beam 30 mm × 30 mm in cross section if the maximum bending moment is 2,200 N.m.
3. Calculate the maximum bending stress on a hollow circular cross section of a beam if the maximum bending moment is 5,000 N.m (Fig. 9.10).

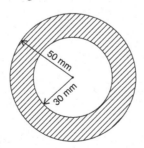

Fig. 9.10

4. Calculate the maximum bending moment on a 6 in. × 14 in. rectangular beam with a maximum bending stress of 15,000 psi.
5. Calculate the maximum bending stress on a beam cross section (Fig. 9.11) with a bending moment of 120,000 lb-in.

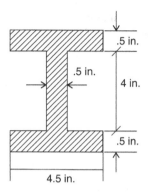

Fig. 9.11

6. Calculate the maximum bending stress on a bar (Fig. 9.12) with a circular cross section of 0.75 in. subjected to a concentrated load of 300 lb.

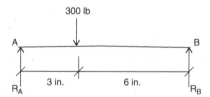

Fig. 9.12

7. Calculate the applied bending moment on a 60 mm × 160 mm rectangular beam with a maximum bending stress of 20,000 kPa.
8. W 10 × 60 simply supported steel beam 20 ft in length carries a uniformly distributed load of 1,500 lb/ft. Determine the maximum bending stress.

Chapter Summary

When a beam is subjected to external loads, shear forces and bending moments develop in the beam. Therefore, a beam must resist these external shear forces and bending moments. The beam itself must develop internal resistance to resist shear forces and bending moments.

The stresses caused by the bending moments are called bending stresses. For beam design purposes, it is very important to calculate the shear stresses and bending stresses at various locations of a beam. The bending stress varies from zero at the neutral axis to a maximum at the tensile and compressive side of the beam.

In the following problems, you might need to refer to the tables of properties of materials found in the engineering texts to extract the necessary information.

$$\sigma = Mc/I$$

where M is the maximum bending moment on the beam; c is the distance from the neutral axis to the outside surface of the beam, and I is the moment of inertia of the area of the beam cross section.

Review Questions

1. What is the *flexure formula*?
2. What is the *neutral axis*?
3. What is the *neutral plane*?
4. How would you calculate bending stress using the flexure formula?
5. How would you calculate the bending moment using the flexure formula?
6. What are the limitations of the flexure formula?
7. What is the *section modulus*?
8. What is the physical meaning of section modulus?
9. How would you calculate the section modulus of a rectangular beam?
10. What is the value of the section modulus in a beam, given $M = 150,000$ lb-in. and bending stress $\sigma = 1,200$ psi?
11. What is the moment inertia of a rectangular beam with cross section of $b = 4$ in. and $h = 9$ in.?
12. Given the section modulus of a beam $S = 80$ in.^{3} and the maximum bending stress $\sigma = 1,500$ psi, what is the value of the maximum bending moment?

Problems

In the following problems, you might need to refer to the tables of properties of materials found in the engineering texts to extract the necessary information.

1. A square steel bar 40 mm on each side is subjected to a bending moment of 550 N.m. Determine the maximum bending moment.
2. Calculate the maximum bending moment for a S 10 × 35 beam, if the bending stress is 18,000 psi.
3. Calculate the maximum bending stress for a W 10 × 33 beam, 12 ft length, if it is subjected to a uniformly distributed load of 1,000 lb/ft.

4. Calculate the maximum bending stress on an 8-in. wide and 18-in. deep beam with a span of 20 ft (Fig. 9.13) at 7 ft from the left.

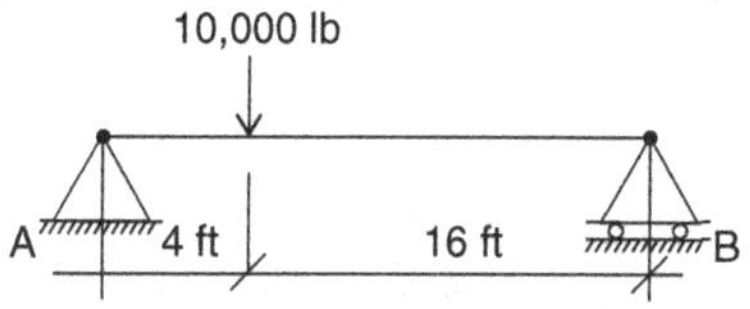
10,000 lb
A
4 ft
16 ft
B

Fig. 9.13

5. Calculate the maximum bending stress for a beam with a cross section of 80 mm × 200 mm. Assume the developed maximum bending moment is 2,000 N.m.
6. A steel bar of 0.7 × 1.8 in. rectangular cross section creates a bending stress of $\sigma = 30{,}000$ psi. Calculate the maximum bending moment developed on the beam.
7. Calculate the maximum tensile and compressive bending stress for the beam shown below (Fig. 9.14) if the applied bending moment is 110,000 N.m.

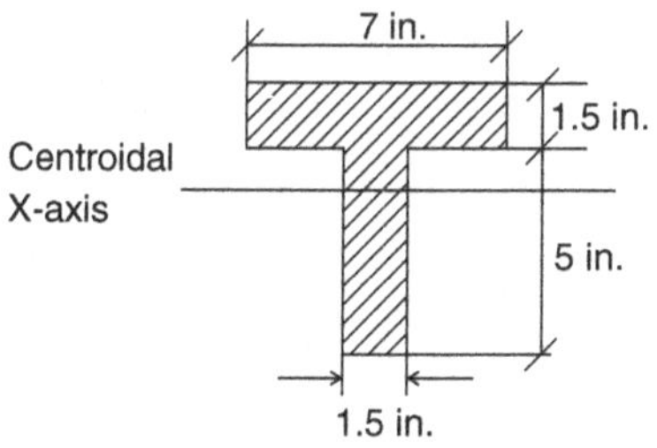
7 in.
1.5 in.
Centroidal
X-axis
5 in.
1.5 in.

Fig. 9.14

8. Calculate the maximum bending stress for a hollow steel pipe with the dimensions shown (Fig. 9.15) subjected to a concentrated load of 30,000 lb. Assume the span of the pipe is 40 in. (OD = 5.5 in., ID = 3.0 in.).

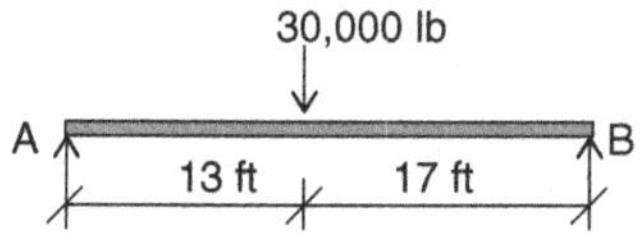
30,000 lb
A
B
13 ft
17 ft

Fig. 9.15

9. Calculate the maximum bending stress for a round steel rod 20 mm in diameter if it is subjected to an applied bending moment of 320 N.m.

10. For the I-section beam shown (Fig. 9.16), calculate the maximum tensile and compressive stress if the beam is acted upon by a bending moment of 100,000 lb-ft.

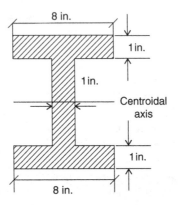

Fig. 9.16

Columns

10

Overview

Structural members that are subjected to axial compressive loads are called columns; their lengths are several times greater than their cross-sectional dimensions. Columns have wide application in the construction industry and are very important in overall design for buildings and bridges. Columns can bend under compressive axial loading, and they might be subject to eccentric axial loading. This phenomenon is called buckling; this means that the column has lost its stability by buckling. In this chapter the buckling of a column as a result of bending action due to axial compressive loads will be discussed.

Learning Objectives

Upon completion of this chapter, you will be able to define columns and calculate the slenderness ratio for timber and steel columns. You will also be able to calculate the allowable load for timber and steel columns. Your knowledge, application, and problem solving skill will be determined by your performance on the chapter test.

10.1 Introduction

A column is a structural member subjected to an end load whose action line parallels that of the member and whose length is generally ten or more times its least lateral dimension. The most common materials for columns are steel, wood, aluminum, and concrete. Normally, the stronger the material, the larger the load the column can carry. Shorter columns have more load-carrying capacity than longer columns.

Generally, all axially loaded columns are called compressive members. There are other structural members that carry compressive loads, but they are not necessarily columns, such as pillars and piers. The size of the column is another factor that has an effect on the load-carrying capacity of the column. End connections of the column have an effect on the load-carrying capacity of the column. End

© Springer International Publishing Switzerland 2015
P. Ghavami, *Mechanics of Materials*, DOI 10.1007/978-3-319-07572-3_10

connections of a column are classified as pinned and fixed connections (Fig. 10.1). Columns with fixed end connections have higher load-carrying capacities.

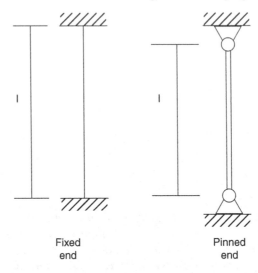

Fixed
end

Pinned
end

Fig. 10.1

10.2   Slenderness Ratio

The slenderness ratio is a very important quantity in the design process of compression members. The slenderness ratio is equal to the length of the compression member divided by the least radius of gyration of the cross section. It is shown by L/r, where L is length of the member and r is the least radius of gyration of the cross section.

It is written as

$$r = \sqrt{I/A}$$

In order to get the least radius of gyration, we have to choose the smallest moment of inertia, either I_x or I_y in the formula.

For low-carbon steel, a short compression member is one that has a slenderness ratio below about 40; for aluminum–magnesium alloys, any compression member that has a slenderness ratio below 10 is considered a short compression member.

Example 10.1 Calculate the slenderness ratio of a compression member 8 ft long with the cross section shown (Fig. 10.2).

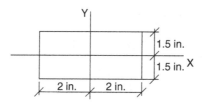

Fig. 10.2

Solution The formula for the slenderness ratio is

$$r = \sqrt{I/A}$$

The moment of inertia of the cross section with respect to the x axis is

$$I_x = (1/12)\,bh^3 = (1/12)(4)(3)^2 = 9\,\text{in.}^4$$

The moment of inertia of the cross section with respect to the y axis is

$$I_y = (1/12)\,bh^3 = (1/12)(3)(4)^2 = 16\,\text{in.}^4$$

The area of the rectangular shape is

$$A = bh = 4 \times 3 = 12\,\text{in.}^2$$

We use the smallest value of moment of inertia in the calculation of the radius of gyration

$$r = \sqrt{I/A} = \sqrt{9/12} = 0.87\,\text{in.}$$

And the slenderness ratio is $L/r = (8 \times 12)/0.87 = 111$

Example 10.2 Calculate the slenderness ratio of a solid circular bar 30 in. long and 2 in. in diameter.

Solution The formula for the slenderness ratio is

$$r = \sqrt{I/A}$$

The moment of inertia of the circular area is the same with respect to the x axis and the y axis.

$$I_x = I_y = \pi/64(d)^4 = \pi/64(2)^4 = 0.79\,\text{in.}^4$$

The area of the rectangular shape is

$$A = \pi d^2/4 = 3.14\,\text{in.}^2$$

and

$$r \qquad = \sqrt{I/A}$$
$$= \sqrt{0.79/3.14} = 0.502\,\text{in.}$$

The slenderness ratio is $L/r = (40)/0.502 = 79.75$

Example 10.3 Calculate the slenderness ratio of an angle 1,000 mm long designated as $64 \times 64 \times 6.4$ mm.

Solution Referring to the tables of properties for angles, we find that $r_x = r_y = 19.5$ mm
And the slenderness ratio is

$$L/r = 1,000/19.5 = 51.3$$

Practice Problems

1. Calculate the slenderness ratio of a rectangular bar 2 in. $\times$ 75 in. and 20 in. long.
2. Calculate the slenderness ratio of a solid circular bar 30 in. long with a diameter of $1\frac{1}{2}$ in.
3. Calculate the slenderness ratio of an angle 750 mm long designated as $51 \times 52 \times 9.6$ mm.
4. Calculate the slenderness ratio of a W-shaped steel beam 20 m long designated as W 300×4.90 mm.
5. Calculate the slenderness ratio of a hollow circular section of a compression member 5 m long (Fig. 10.3).

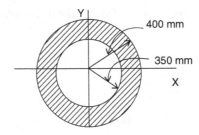

Fig. 10.3

10.3   Failure in Columns

Columns can be classified as three types: short, long, and intermediate (Fig. 10.4). Failure in short columns is distinguished by the crushing (yielding) of material, due to the stresses developed in the material which are greater than the yielding stress. Long column fails are due to buckling, meaning that the column actually defects laterally and bends near the middle of its length. Buckling in long columns occurs at stresses below the yielding stress of the material. Failure in intermediate columns is a combination of material yielding and buckling. Buckling in intermediate columns takes place at higher stresses.

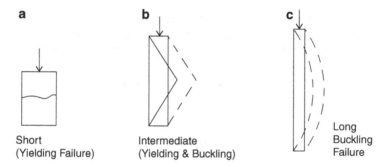

Fig. 10.4

Leonhard Euler (1707–1783), Swiss mathematician, derived a formula for a slender column that shows how the critical load would cause buckling in this type of column. The formula is the Euler equation and was named after him for his great contribution to column theory. Today it is still used as the basis for the analysis and design of slender columns. The Euler equation is expressed as

$$P_e = \pi^2 EI/L^2 \qquad (10.1)$$

where

P_e = the critical load or the Euler buckling load, lb or N.
E = the modulus of elasticity of the material, psi or MPa
I = the smallest moment of inertia of the cross section, in.^{4} or mm^4
L = the length of the column between the hinges, in. or mm

The Euler formula is used to precisely predict the buckling load if the buckling stress is less than the proportional limit of the material. For the purpose of comparison, the Euler formula can be expressed in terms of buckling stress using Eq. 10.1 and substituting Ar^2 for the moment of inertia, I.

$$P_e/A = \left(\pi^2 EI\right)/(L/r)^2$$

or

$$\sigma_e = \left(\pi^2 E\right)/(L/r)^2 \tag{10.2}$$

The Euler equation has some limitations in the fact that it is only used if buckling occurs, as stresses on a column are low and in the material's elastic region.

Example 10.4 Calculate the critical load using the Euler formula for a 250-mm long rectangular steel bar 15 mm by 4 mm, to be used as a pin-connected compression member. Assume the modulus of elasticity for steel is 200×10^3 MPa.

Solution First, we must find out if the Euler formula can be used for this problem. In other words, the compression member must be a slender column.
Calculate the moment of inertia I_x and I_y to find out which one is smaller.

$$I_x = 1/12(15)(4)^3 = 80\,\text{mm}^4$$

and

$$I_y = 1/12(4)(15)^3 = 1,124\,\text{mm}^4$$

I_x is smaller than I_y.
The area is $A = 15 \times 4 = 60\,\text{mm}^2$
The radius of gyration is

$$
\begin{aligned}
r &= \sqrt{I/A} \\
&= \sqrt{80/60} = 1.15\,\text{mm}
\end{aligned}
$$

The slenderness ratio is

$$L/r = 250\,\text{mm}/1.15\,\text{mm} = 217$$

Since $217 > 150$ (steel), the compression member is a slender column, and the Euler formula can be applied.

$$
\begin{aligned}
P_e &= \pi^2 EI/L^2 \\
&= \left[\pi^2 \times 200 \times 10^3 \times 80\right]/(250)^2 = 2,527\,\text{N}
\end{aligned}
$$

Example 10.5 Calculate the smallest value of the slenderness ratio for a given column with a modulus of elasticity of 15×10^6 psi and critical buckling stress of 20,000 psi so that the Euler equation can be applied with some accuracy.

Solution   The buckling stress equation is written as

$$\sigma_e = \left(\pi^2 E\right)/(L/r)^2$$

or

$$(L/r)^2 = \left(\pi^2 E\right)/\sigma_e$$

$$L/r = \pi\sqrt{E}/\sigma_e$$
$$= \pi\left(\sqrt{15 \times 10^6/20,000}\right) = 28.5$$

Practice Problems

1. Calculate the critical Euler buckling stress for a W 12×40 steel column 24 ft in length with a modulus of elasticity of 3×10^7 psi.
2. Using the Euler formula, calculate the critical load of a 1.5-in. diameter steel rod 3 ft long used as a compression member. Assume end conditions are pinned-connected and the modulus of elasticity is 3×10^7 psi.
3. Using the Euler formula, calculate the critical buckling stress for a W-shaped 16×26 steel column 16 ft long with a modulus of elasticity of 3×10^7 psi.
4. Calculate the critical Euler buckling stress for a steel pipe column 6 in. in diameter and 20 ft long. The proportional limit for steel is 34,000 psi.
5. Calculate the critical Euler buckling stress for a steel rod ¾ in. in diameter and 6 ft long. The proportional limit for steel is 34,000 psi.

10.4   Timber Columns

Simple, solid, square or rectangular wood columns are the most frequently used types of timber in the construction industry today. Round cross-sectional columns are also classified as simple solid columns, but are used less frequently. The other types of timber columns are spaced columns and built-up columns.

In this section, we only discuss the simple, solid square or rectangular cross section. The most widely used design code for the analysis and design of timber columns is recommended by the *National Forest Products Association* (NFPA), under the National Design Specification for Wood Construction.

10.5   Allowable Stress in Timber Columns

The long column (slender column) is defined by the L/d ratio, where L is the actual length of the column, and d is the smallest dimension of the column cross section. The slenderness ratio for a simple solid timber column is limited to $L/d = 50$.

For timber columns, the allowable axial compressive stress is given by the formula

$$\sigma_{\text{all}} = 0.3E/(L/d)^2$$

where

E = the modulus of elasticity of timber, psi
L = actual length of the column, in.
d = the least dimension of the column, in.

The above equation is basically the same as the Euler equation, except for the new definition of L/d for a timber column. This ratio is limited to 50 or less.

Example 10.6 Determine the allowable axial compressive stress for a pin-connected 4×8 timber column of Douglas fir that is 10 ft long. The table value of allowable stress is ($s_c = 1,050$ psi).

Solution
$$L/d = 10(12)/3.5 = 34.3 < 50 \quad \textbf{OK}$$

[Note that the dressed value of 4×8 Douglas structural timber is $3\frac{1}{2} \times 7\frac{1}{4}$.]
Now we must calculate a factor called K to find out if the column can perform as a slender one.

$$K = 0.671\sqrt{E/s_c} = 0.671\sqrt{1,700,000/1,050} = 27$$

For longer columns ($K \leq L/d \leq 50$)
Since $27 < 34.3 < 50$, it is a slender column and we use

$$\begin{aligned}\sigma_{\text{all}} &= 0.3E/(L/d)^2 \\ &= 0.3(1,700,000)/(34.3)^2 \\ &= 433\,\text{psi} < 1,050\,\text{psi} \quad \textbf{OK}\end{aligned}$$

Example 10.7 Determine the allowable axial compressive load for a pin-connected 6×6 timber column of Hem-fir that is 20 ft long. The table value of allowable stress is ($s_c = 875$ psi). $E = 1,400,000$ psi.

Solution
$$L/d = 20 \times 12/5.5 = 43.6 < 50 \quad \textbf{OK}$$

Now we must calculate a factor called K to find out if the column can perform as a slender one.

$$K = 0.671\sqrt{E/s_c} = 0.671\sqrt{1,400,000/875} = 26.8$$

For longer columns ($K \leq L/d \leq 50$)

Since $26.8 < 43.6 < 50$, it is a slender column and we use

$$\sigma_{all} = 0.3E/(L/d)^2$$
$$= 0.3(1,400,000)/(43.6)^2$$
$$= 221\,psi < 875\,psi$$

Chapter Summary

Here is a brief summary of what you have learned in this chapter.

Columns are generally considered as compression members and classified as short, intermediate, and long (slender). Columns fail due to a type of bending called buckling. In short and intermediate columns, the buckling occurs due to inelastic bending. In long columns, the buckling occurs due to an elastic bending.

The Euler equation is used for calculating the buckling critical load and is expressed as

$$P_e = \pi^2 EI/L^2$$

The Euler formula is limited to the proportional limit of the material.
The critical stress is

$$\sigma_e = \left(\pi^2 E\right)/(L/r)^2$$

L/r is called the slenderness ratio. For a steel slender column, the slenderness ratio is above 150.

For timber columns, the allowable stress based on the Euler formula is given as

$$\sigma_{all} = 0.3E/(L/d)^2$$

The codes limit L/d to 50 or less.

Review Questions

1. What is a *compression member?*
2. What is the *radius of gyration?*
3. What is the slenderness ratio and its importance in column issues?
4. What are the compression member categories?
5. How does a short compression member fail?
6. How does an intermediate compression member fail?
7. What is the Euler formula, its applications, and limitations?
8. What is the critical load for steel slender compression members?
9. What formula would you use to calculate the allowable stress for timber columns?
10. What is the ratio L/d in timber columns?

Problems

1. What is the slenderness ratio of a solid circular bar 30 in. long and 2 in. in diameter?
2. Calculate the slenderness ratio of a rectangular bar 1.5 in. × ¾ in. and 20 in. long.
3. Calculate the compressive stress of a hollow circular machine part ¾ in. long and made of steel with an outer diameter of 4 in. and an inner diameter of 2 in. Use 80,000 lb as a compressive force.
4. What is the slenderness ratio of a steel beam 12 ft long and designated as W 10 × 12?
5. Calculate the critical Euler buckling stress for a W 12 × 58 steel column 20 ft in length with a modulus of elasticity of 3×10^7 psi.
6. Using the Euler formula, calculate the critical load of a 2-in. diameter steel rod 4 ft long used as a compression member. Assume end conditions are pinned-connected and the modulus of elasticity is 3×10^7 psi.
7. Using the Euler formula, calculate the critical buckling stress for a W 14 × 22 shape steel column 12 ft long with a modulus of elasticity of 3×10^7 psi.
8. Calculate the critical Euler buckling stress for a steel pipe column of 5 in. outer diameter, 3 in. inner diameter, and 15 ft long. The proportional limit for steel is 34,000 psi.
9. Calculate the critical Euler buckling stress for a steel rod ½-in. in diameter and 8 ft long. The proportional limit for steel is 34,000 psi.
10. Determine the allowable axial compressive stress for a pin-connected 4 × 10 timber column of Douglas fir that is 12 ft long. The table value of allowable stress is ($s_c = 1{,}050$ psi).

Appendix A. Shears, Moments and Deflections

1. Simple beam—uniformly distributed load

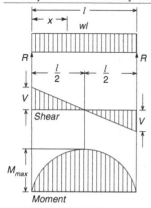

	Total Equiv. Uniform Load = wl
	$R = V = \dfrac{wl}{2}$
	$V_x = w\left(\dfrac{1}{2} - x\right)$
	$M_{max}\,(\text{at center}) = \dfrac{wl^2}{8}$
	$M_x = \dfrac{wx}{2}(l - x)$
	$\Delta_{max}\,(\text{at center}) = \dfrac{5wl^4}{384\,El}$
	$\Delta_x = \dfrac{wx}{24\,El}(l^3 - 2lx^2 + x^3)$

2. Simple beam—load increasing uniformly to one end

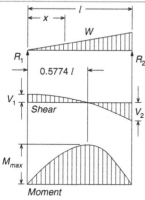

	Total Equiv. Uniform Load
	$= \dfrac{16W}{9\sqrt{3}} = 1.03\,W$
	$R_1 = V_1 = \dfrac{W}{3}$
	$R_2 = V_2 = V_{max} = \dfrac{2W}{3}$
	$V_x = \dfrac{W}{3} - \dfrac{Wx^2}{l^2}$
	$M_{max}\left(\text{at } x = \dfrac{1}{\sqrt{3}} = 0.557\,l\right)$
	$= \dfrac{2Wl}{9\sqrt{3}} = 0.128\,Wl$
	$M_x = \dfrac{Wx}{3l^2}(l^2 - x^2)$
	$\Delta_{max}\left(\text{at } X = l\sqrt{1 - \sqrt{\dfrac{8}{15}}} = 0.519\,l\right) = 0.0130\dfrac{Wl^3}{El}$
	$\Delta_x = \dfrac{WX}{180\,Ell^2}(3X^4 - 10l^2X^2 + 7l^4)$

(continued)

© Springer International Publishing Switzerland 2015
P. Ghavami, *Mechanics of Materials*, DOI 10.1007/978-3-319-07572-3

(continued)

---

3. Simple beam—load increasing uniformly to center

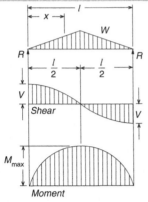

$$\text{Total Equiv. Uniform Load} = \frac{4W}{3}$$

$$R = V = \frac{W}{2}$$

$$V_X\left(\text{when }X < \frac{l}{2}\right) = \frac{W}{2l^2}(l^2 - 4X^2)$$

$$M_{\max}(\text{at center}) = \frac{Wl}{6}$$

$$M_X\left(\text{when }X < \frac{l}{2}\right) = WX\left(\frac{1}{2} - \frac{2X^2}{3l^2}\right)$$

$$\Delta_{\max}(\text{at center}) = \frac{Wl^2}{60EI}$$

$$\Delta_x\left(\text{when }x < \frac{l}{2}\right) = \frac{wx}{480EIl^2}\left(5l^2 - 4x^2\right)^2$$

---

4. Simple beam—uniform load partially distributed

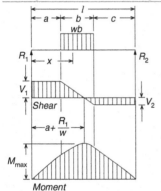

$$R_1 = V_1(\text{max. when }a < c) = \frac{wb}{2l}(2c + b)$$

$$R_2 = V_2(\text{max. when }a > c) = \frac{wb}{2l}(2a + b)$$

$$V_x(\text{when }x > a \text{ and} < (a+b)) = R_1 - w(x - a)$$

$$M_{\max}\left(\text{at }x = a + \frac{R_1}{w}\right) = R_1\left(a + \frac{R_1}{2w}\right)$$

$$M_x(\text{when }x < a) = R_1x$$

$$M_x(\text{when }x > a \text{ and} < (a+b))$$

$$= R_1x - \frac{w}{2}(x - a)^2$$

$$M_x(\text{when }x > (a+b)) = R_2(l - x)$$

(continued)

(continued)

---

5. Simple beam—uniform load partially distributed at one end

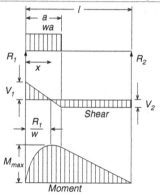

$$R_1 = V_1 = V_{max} = \frac{wa}{2l}(2l - a)$$

$$R_2 = V_2 = \frac{wa^2}{2l}$$

$$V_x(\text{when } x < a) = R_1 - wx$$

$$M_{max}\left(\text{at } x = \frac{R_1}{w}\right) = \frac{R_1^2}{2w}$$

$$M_x(\text{when } x < a) = R_1 x - \frac{wx^2}{2}$$

$$M_x(\text{when } x > a) = R_2(l - x)$$

$$\Delta_x(\text{when } x < a) = \frac{wx}{24 EIl}$$
$$\left(a^2(2l - a)^2 - 2ax^2(2l - a) + lx^3\right)$$

$$\Delta_x(\text{when } x > a) = \frac{wa^2(l - x)}{24 EIl}$$
$$\left(4xl - 2x^2 - a^2\right)$$

---

6. Simple beam—uniform load partially distributed at each end

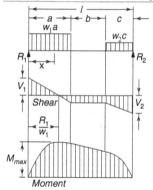

$$R_1 = V_1 = \frac{w_1 a(2l - a) + w_2 c^2}{2l}$$

$$R_2 = V_2 = \frac{w_2 c(2l - c) + w_1 a^2}{2l}$$

$$V_x(\text{when } x < a) = R_1 - w_1 x$$

$$V_x(\text{when } a < x < (a + b)) = R_1 - w_1 a$$

$$V_x(\text{when } x > (a + b)) = R_2 + w_2(l - x)$$

$$M_{max}\left(\text{at } x = \frac{R_1}{w_1}, \text{when } R_1 < w_1 a\right) = \frac{R_1^2}{2w_1}$$

$$M_{max}\left(\text{at } x = l - \frac{R_2}{w_2}, \text{when } R_2 < w_2 c\right) = \frac{R_2^2}{2w_2}$$

$$M_x(\text{when } x < a) = R_1 x - \frac{w_1 x^2}{2}$$

$$M_x(\text{when } a < x < (a + b)) = R_1 x - \frac{w_1 a}{2}(2x - a)$$

$$M_x(\text{when } x > (a + b)) = R_2(l - x) - \frac{w_2(l - x)^2}{2}$$

(continued)

(continued)

7. Simple beam—concentrated load at center

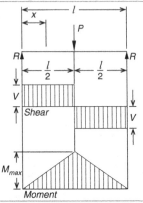

Total Equiv. Uniform Load $= 2P$

$R = V = 2P$

M_{max} (at point of load) $= \dfrac{Pl}{4}$

$M_x \left(\text{when} \, x < \dfrac{1}{2} \right) = \dfrac{Px}{2}$

Δ_{max} (at point of load) $= \dfrac{Pl^3}{48 \, EI}$

$\Delta_x \left(\text{when} \, x < \dfrac{1}{2} \right) = \dfrac{Px}{48 \, EI} \left(3l^2 - 4x^2 \right)$

8. Simple beam—concentrated load at any point

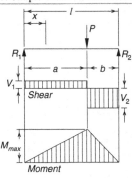

Total Equiv. Uniform Load $= \dfrac{8Pab}{l^2}$

$R_1 = V_1 (= V_{max} \, \text{when} \, a < b) = \dfrac{Pb}{l}$

$R_2 = V_2 (= V_{max} \, \text{when} \, a > b) = \dfrac{Pa}{l}$

M_{max} (at point of load) $= \dfrac{Pab}{l}$

$M_x (\text{when} \, x < a) = \dfrac{Pbx}{l}$

$\Delta_{max} \left(\text{at} \, x = \sqrt{\dfrac{a(a+2b)}{3}}, \ \text{when} \ a > b \right)$

$= \dfrac{Pab(a+2b)\sqrt{3a(a+2b)}}{27EIl}$

Δ_a (at point of load) $= \dfrac{Pa^2b^2}{3EIl}$

$\Delta_x (\text{when} \, x < a) = \dfrac{Pbx}{6EIl} \left(l^2 - b^2 - x^2 \right)$

9. Simple beam—two equal loads symmetrically placed

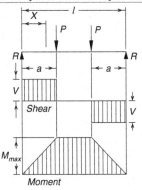

Total Equiv. Uniform Load $= \dfrac{8Pa}{l}$

$R = V = P$

$M_{max} (\text{between loads}) = Pa$

$M_x (\text{when} \, x < a) = Px$

Δ_{max} (at center) $= \dfrac{Pa}{24 EI} \left(3l^2 - 4a^2 \right)$

$\Delta_{max} \left(\text{when} \, a = \tfrac{l}{3} \right) = \dfrac{Pl^3}{28 EI}$

$\Delta_x (\text{when} \, x < a) = \dfrac{Px}{6EI} \left(3la - 3a^2 - x^2 \right)$

$\Delta_x (\text{when} \, a < x < (l-a)) = \dfrac{Pa}{6EI} \left(3lx - 3x^2 - a^2 \right)$

(continued)

(continued)

---

10. Simple beam—two equal concentrated loads unsymmetrically placed

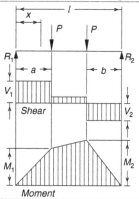

$$R_1 = V_1 \left(= V_{\max} \text{ when } a < b \right) = \frac{P}{l}(l - a + b)$$

$$R_2 = V_2 \left(= V_{\max} \text{ when } a > b \right) = \frac{P}{l}(l - a + b)$$

$$V_x \left(\text{when } a < x < (l - b) \right) = \frac{P}{l}(b - a)$$

$$M_1 \left(= M_{\max} \text{ when } a > b \right) = R_1 a$$

$$M_2 \left(= M_{\max} \text{ when } a > b \right) = R_2 b$$

$$M_x \left(\text{when } x < a \right) = R_1 x$$

$$M_x \left(\text{when } a < x < (l - b) \right) = R_1 x - P(x - a)$$

---

11. Simple beam—two unequal concentrated loads unsymmetrically placed

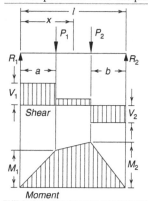

$$R_1 = V_1 = \frac{P_1(l - a) + P_2 b}{l}$$

$$R_2 = V_2 = \frac{P_1 a + P_2(l - b)}{l}$$

$$V_x (\text{when } a < x < (l - b)) = R_1 - P_1$$

$$M_1 (= M_{\max} \text{ when } R_1 < P_1) = R_1 a$$

$$M_2 (= M_{\max} \text{ when } R_2 < P_2) = R_2 b$$

$$M_x (\text{when } x < a) = R_1 x$$

$$M_x (\text{when } a < x < (l - b)) = R_1 x - P_1(x - a)$$

---

12. Beam fixed at one end, supported at other—uniformly distrubted load

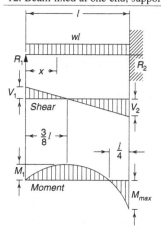

Total Equiv. Uniform Load $= wl$

$$R_1 = V_1 = \frac{3wl}{8}$$

$$R_2 = V_2 = V_{\max} = \frac{5wl}{8}$$

$$V_x = R_1 - wx$$

$$M_{\max} = \frac{wl^2}{8}$$

$$M_1 \left(\text{at } x = \frac{3}{8}l \right) = \frac{9}{128}wl^2$$

$$M_x = R_1 x - \frac{wx^2}{2}$$

$$\Delta_{\max} \left(\text{at } x = \frac{l}{16}(1 + \sqrt{33}) = 0.422\, l \right) = \frac{wl^4}{185 EI}$$

$$\Delta_x = \frac{wx}{48 EI}(l^3 - 3lx^2 + 2x^3)$$

(continued)

13. Beam fixed at both ends—uniformly distributed loads

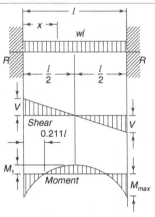

$$\text{Total Equiv. Uniform Load} = \frac{2wl}{3}$$

$$R = V = \frac{wl}{2}$$

$$V_x = w\left(\tfrac{l}{2} - x\right)$$

$$M_{\max}(\text{at ends}) = \frac{wl^2}{12}$$

$$M_1(\text{at center}) = \frac{wl^2}{24}$$

$$M_x = \frac{w}{12}\left(6lx - l^2 - 6x^2\right)$$

$$\Delta_{\max}(\text{at center}) = \frac{wl^4}{384EI}$$

$$\Delta_x = \frac{wx^2}{24EI}(l - x)^2$$

14. Beam fixed at both ends—concentrated load at center

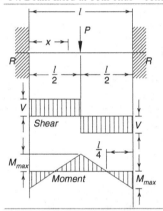

$$\text{Total Equiv. Uniform Load} = P$$

$$R = V = \frac{P}{2}$$

$$M_{\max}(\text{at center and ends}) = \frac{Pl}{8}$$

$$M_x\left(\text{when } x < \frac{l}{2}\right) = \frac{P}{8}(4x - 1)$$

$$\Delta_{\max}(\text{at center}) = \frac{Pl^3}{192EI}$$

$$\Delta_x\left(\text{when } x < \frac{l}{2}\right) = \frac{Px^2}{48EI}(3l - 4x)$$

15. Beam fixed at both ends—concentrated load at any point

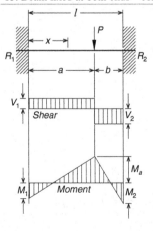

$$R_1 = V_1\left(= V_{\max} \text{ when } a < b\right) = \frac{Pb^2}{l^3}(3a + b)$$

$$R_2 = V_2\left(= V_{\max} \text{ when } a > b\right) = \frac{Pb^2}{l^3}(a + 3b)$$

$$M_1\left(= M_{\max} \text{ when } a < b\right) = \frac{Pab^2}{l^2}$$

$$M_2\left(= M_{\max} \text{ when } a > b\right) = \frac{Pa^2b}{l^2}$$

$$M_a(\text{at point of load}) = \frac{2Pa^2b^2}{l^3}$$

$$M_x(\text{when } x < a) = R_1x - \frac{Pab^2}{l^2}$$

$$\Delta_{\max}\left(\text{when } a > b \text{ at } x = \frac{2al}{3a + b}\right) = \frac{2Pa^3b^2}{3EI(3a + b)^2}$$

$$\Delta_a(\text{at point of load}) = \frac{Pa^3b^3}{3EIl^3}$$

$$\Delta_x(\text{when } x < a) = \frac{Pb^2x^2}{6EIl^3}(3al - 3ax - bx)$$

(continued)

16. Cantilevered beam—load increasing uniformly to fixed end

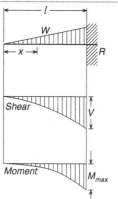

Total Equiv. Uniform Load $= \dfrac{8}{3}W$

$R = V = W$

$V_x = W\dfrac{x^2}{l^2}$

$M_{max}\,(\text{at fixed end}) = \dfrac{Wl}{3}$

$M_x = \dfrac{Wx^3}{3l^2}$

$\Delta_{max}\,(\text{at free end}) = \dfrac{Wl^3}{15EI}$

$\Delta_x = \dfrac{W}{60EIl^2}\left(x^5 - 5l^4x + 4l^5\right)$

17. Cantilevered beam—uniformly distributed load

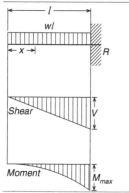

Total Equiv. Uniform Load $= 4\,wl$

$R = V = wl$

$V_x = wx$

$M_{max}\,(\text{at fixed end}) = \dfrac{wl^2}{2}$

$M_x = \dfrac{wx^{\frac{3}{2}}}{2}$

$\Delta_{max}\,(\text{at free end}) = \dfrac{wl^4}{8EI}$

$\Delta_x = \dfrac{w}{24EI}\left(x^4 - 4l^3x + 3l^4\right)$

18. Beam fixed at one end, free to deflect vertically but not rotate at other—uniformly distributed load

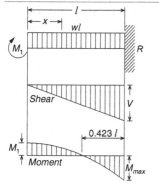

Total Equiv. Uniform Load $= \dfrac{8}{3}wl$

$R = V = wl$

$V_x = wx$

$M_1\,(\text{at deflected end}) = \dfrac{wl^2}{6}$

$M_{max}\,(\text{at fixed end}) = \dfrac{wl^2}{3}$

$M_x = \dfrac{w}{6}\left(l^2 - 3x^2\right)$

$\Delta_{max}\,(\text{at deflected end}) = \dfrac{wl^4}{24EI}$

$\Delta_x = \dfrac{w\left(l^2 - x^2\right)^2}{24EI}$

(continued)

(continued)

19. Cantilevered beam—concentrated load at any point

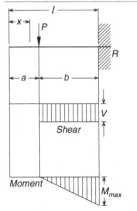

Total Equiv. Uniform Load $= \dfrac{8Pb}{L}$
$R = V = P$
$M_{\max}$ (at fixed end) $= Pb$
M_x (when $x > a$) $= P(x - a)$
$\Delta_{\max}$ (at free end) $= \dfrac{Pb^2}{6EI}(3l - b)$
Δ_a (at point of load) $= \dfrac{Pb^3}{3EI}$
Δ_x (when $x < a$) $= \dfrac{Pb^2}{6EI}(3l - 3x - b)$
Δ_x (when $x > a$) $= \dfrac{P(l - x)^2}{6EI}(3b - l - x)$

20. Cantilevered beam—concentrated load at free end

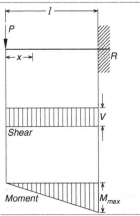

Total Equiv. Uniform Load $= 8P$
$R = V = P$
$M_{\max}$ (at fixed end) $= Pl$
$M_x = Px$
$\Delta_{\max}$ (at free end) $= \dfrac{Pl^3}{3EI}$
$\Delta_x = \dfrac{P}{6EI}(2l^3 - 3lx^2 + x^3)$

21. Beam fixed at one end, free to deflect vertically but not rotate at other—concentrated load at deflected end

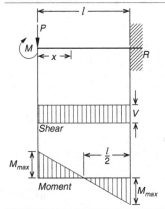

Total Equiv. Uniform Load $= 4P$
$R = V = P$
$M_{\max}$ (at both ends) $= \dfrac{Pl}{2}$
$M_x = P\left(\dfrac{l}{2} - x\right)$
$\Delta_{\max}$ (at deflected end) $= \dfrac{Pl^3}{12EI}$
$\Delta_x = \dfrac{P(l - x)^2}{12EI}(l + 2x)$

Reproduced courtesy of the American Institute of Steel Construction

Appendix B. Centroids and Properties of Areas

Table B.1

Figure	Area	Location Of centroid	Moment of Inertia I_x	Section modulus S_x
	bh	$\bar{y} = \dfrac{h}{2}$	$\dfrac{bh^3}{12}$	$\dfrac{bh^2}{6}$
	a^2	$\bar{y} = \dfrac{a}{2}$	$\dfrac{a^4}{12}$	$\dfrac{a^3}{6}$
	$\dfrac{bh}{2}$	$\bar{y}_1 = \dfrac{h}{3}$ $\bar{y}_2 = \dfrac{2h}{3}$	$\dfrac{bh^3}{36}$	$s_1 = \dfrac{I_x}{\bar{y}_1} = \dfrac{bh^2}{12}$ $s_2 = \dfrac{I_x}{\bar{y}_2} = \dfrac{bh^2}{24}$

(continued)

© Springer International Publishing Switzerland 2015
P. Ghavami, *Mechanics of Materials*, DOI 10.1007/978-3-319-07572-3

Table B.1 (continued)

Figure	Area	Location Of centroid	Moment of Inertia I_x	Section modulus S_x
	$\dfrac{bh}{2}$	$\bar{y}_1 = \dfrac{h}{3}$ $\bar{y}_2 = \dfrac{2h}{3}$	$\dfrac{bh^3}{36}$	$s_1 = \dfrac{bh^2}{12}$ $s_2 = \dfrac{bh^2}{24}$
	πr^2 or $\dfrac{\pi d^2}{4}$	$\bar{y} = r$ or $\bar{y} = \dfrac{d}{2}$	$\dfrac{\pi r^4}{4}$ or $\dfrac{\pi d^4}{64}$	$\dfrac{\pi r^3}{4}$ or $\dfrac{\pi d^3}{32}$
	$\dfrac{\pi r^2}{2}$	$\bar{y} = r$	$\dfrac{\pi r^4}{8}$	$\dfrac{\pi r^3}{8}$
	$\dfrac{\pi r^2}{2}$	$\bar{y} = 0.05756r$ $\bar{y} = 0.4244r$	$0.1098r^4$	$s_1 = 0.1907r^3$ $s_2 = 0.25886r^3$
	$\dfrac{\pi r^3}{4}$	$\bar{y} = 0.05756r$ $\bar{y} = 0.4244r$	$0.0649r^4$	$s_1 = 0.0953r^3$ $s_2 = 0.1293r^3$

(continued)

Table B.1 (continued)

Figure	Area	Location Of centroid	Moment of Inertia I_x	Section modulus S_x
	$\frac{\pi}{4}(D^2 - d^2)$	$\bar{y} = \frac{D}{2}$	$\frac{\pi D^3 t^a}{8}$	$\frac{\pi D^2}{4}$

^{a}This formula holds for t much smaller than D

Table B.2 Approximate values of the modulus of elasticity E of typical structural materials

Material	Modulus of elasticity E	
	Kips/in.^{2}	kN/cm^2
Steel	30,000	20,700
Wrought iron	28,000	19,300
Brass	15,000	10,300
Cast iron	11,000	7,500
Aluminum	10,000	7,000
Concrete (in compression)	3,000–5,000	2,000–3,400
Timber	1,760	1,200
Granite	1,280	880
Limestone	900	600
Brick	400	280
Plexiglass	400	280
Rubber	1.0	0.7

References

American Institute of Steel Construction (AISC) (2011) Steel construction manual, 14 ed., One East Wacker Drive Suite 700, Chicago, IL 60601-1802

Bauld NR (1986) Mechanics of materials. PWS Engineering, Boston, MA

Beer FP, Johnson R (2010) Statics and mechanics of materials. McGraw-Hill, New York

Buchanan RG (1995) Mechanics of materials. HRW, New York

Burns T (2009) Applied statics and strength of materials. Thomson, New York

Cohen IB (1985) The birth of new physics. W.W. Norton, New York

Den Hartog JP (2012) Strength of materials. Dover, New York

Dugas R (1988) A history of mechanics. Dover, New York

Goodman LE, William WH (2001) Statics. Dover, New York

Harris OC (1982) Statics and strength of materials. Prentice Hall, Englewood Cliffs, NJ

Hibbler RC (2013) Mechanics of materials. Prentice Hall, Englewood Cliffs, NJ

Jackson HJ, Harold WG (1983) Statics and strength of materials. McGraw-Hill, New York

Johnson C (2009) Physics. Wiley, New York

Kashani SMH (1999) Strength of materials. University of Tehran, Tehran

Kline M (1972) Mathematical thought from ancient to modern times. Oxford University Press, New York

Ladislav C (1981) Elementary statics and strength of materials. McGraw-Hill, New York

Motz L, Weaver J (1989) The story of physics. Avon, New York

Muvdi BB, Mcnabb JW (1990) Engineering mechanics of materials. Macmillan, New York

Parker H, Ambrose J (2002) Simplified mechanics and strength of materials. Wiley, New York

Popov PE (1973) Introduction to mechanics of solids. Prentice Hall, Englewood Cliffs, NJ

Spiegel L, Limbrunner G (1994) Applied strength of materials. Macmillan, New York

Timoshenko S (1983) Strength of materials. D. Van Nostrand, New York

Tippens EP (2005) Physics. McGraw-Hill, New York

© Springer International Publishing Switzerland 2015
P. Ghavami, *Mechanics of Materials*, DOI 10.1007/978-3-319-07572-3

Printed in the United States
By Bookmasters